HIGH PERFORMANCE COMPUTATIONAL SCIENCE AND ENGINEERING

IFIP – The International Federation for Information Processing

IFIP was founded in 1960 under the auspices of UNESCO, following the First World Computer Congress held in Paris the previous year. An umbrella organization for societies working in information processing, IFIP's aim is two-fold: to support information processing within its member countries and to encourage technology transfer to developing nations. As its mission statement clearly states,

> *IFIP's mission is to be the leading, truly international, apolitical organization which encourages and assists in the development, exploitation and application of information technology for the benefit of all people.*

IFIP is a non-profitmaking organization, run almost solely by 2500 volunteers. It operates through a number of technical committees, which organize events and publications. IFIP's events range from an international congress to local seminars, but the most important are:

- The IFIP World Computer Congress, held every second year;
- Open conferences;
- Working conferences.

The flagship event is the IFIP World Computer Congress, at which both invited and contributed papers are presented. Contributed papers are rigorously refereed and the rejection rate is high.

As with the Congress, participation in the open conferences is open to all and papers may be invited or submitted. Again, submitted papers are stringently refereed.

The working conferences are structured differently. They are usually run by a working group and attendance is small and by invitation only. Their purpose is to create an atmosphere conducive to innovation and development. Refereeing is less rigorous and papers are subjected to extensive group discussion.

Publications arising from IFIP events vary. The papers presented at the IFIP World Computer Congress and at open conferences are published as conference proceedings, while the results of the working conferences are often published as collections of selected and edited papers.

Any national society whose primary activity is in information may apply to become a full member of IFIP, although full membership is restricted to one society per country. Full members are entitled to vote at the annual General Assembly, National societies preferring a less committed involvement may apply for associate or corresponding membership. Associate members enjoy the same benefits as full members, but without voting rights. Corresponding members are not represented in IFIP bodies. Affiliated membership is open to non-national societies, and individual and honorary membership schemes are also offered.

HIGH PERFORMANCE COMPUTATIONAL SCIENCE AND ENGINEERING

IFIP TC5 Workshop on High Performance Computational Science and Engineering (HPCSE), World Computer Congress, August 22-27, 2004, Toulouse, France

Edited by

Michael K. Ng
The University of Hong Kong
Hong Kong

Andrei Doncescu
LAAS-CNRS
France

Laurence T. Yang
St. Francis Xavier University
Canada

Tau Leng
Supermicro Computer Inc.
USA

Library of Congress Cataloging-in-Publication Data

A C.I.P. Catalogue record for this book is available from the Library of Congress.

High Performance Computational Science and Engineering/ Edited by Michael K. Ng, Andrei Doncescu, Laurence T. Yang, Tau Leng

p.cm. (The International Federation for Information Processing)

ISBN 978-1-4419-3684-4 e-ISBN 978-0-387-24049-7 Printed on acid-free paper.

Printed in the United States of America.

9 8 7 6 5 4 3 2 1
springeronline.com

Contents

Contributing Authors vii

Preface ix

Part I Keynote Talk

Exploiting Multiple Levels of Parallelism in Scientific Computing 3
Thomas Rauber, Gudula Rünger

Part II Distributed Computing

A Spares Distributed Memory Capable of Handling Small Cues 23
Ashraf Anwar, Stan Franklin

Towards a Reslistic Performance Model for Networks of Heterogenous Computers 39
Alexey Lastovetsky, John Twamley

Extending ClusterSim with Message-Passing and Distributed Shared Memory Modules 59
Christiane Pousa, Luiz Ramos, Luis F. Goes, Carlos A. Martins

Rendering Complex Scenes on Clusters with Limited Precomputation 79
Gilles Cadet, Sebastian Zambal, Bernard Lecussan

Part III Numerical Computations

The Evaluation of the Aggregate Creation Orders: Smoothed Aggregation Algebraic MultiGrid Method 99
Akihiro Fujii, Akira Nishida, Yoshio Oyanagi

Pinpointing the Real Zeros of Analytic Functions 123
Soufiane Noureddine, Abdelaziz Fellah

Reducing Overhead in Sparse Hypermatrix Cholesky Factorization 143
Jose Herrero, Juan Navarro

Part IV Computational Applications

Parallel Image Analysis of Morphological Yeast Cells 157
Laurent Manyri, Andrei Doncescu, Jacky Desachy, Laurence T. Yang

An Hybrid Approach to Real Complex System Optimization: Application to Satellite Constellation Design 169
Enguerran Grandchamp

Optimal Decision Making for Airline Inventory Control 201
Ioana Bilegan, Sergio Gonzalez-Rojo, Felix Mora-Camino, Carlos Cosenza

Auotmatic Text Classification Using an Artificial Neural Network 215
Rodrigo Fernandes de Mello, Luciano José Senger, Laurence T. Yang

Contributing Authors

Ashraf Anwar (Arab Academy for Science and Technology, Egypt)

Ioana Bilegan (LAAS-CNRS, France)

Gilles Cadet (Supaero, France)

Carlos Cosenza (COPPE/UFRJ, France)

Jacky Desachy (LAAS-CNRS, France)

Andrei Doncescu (LAAS-CNRS, France)

Abdelaziz Fellah (University of Lethbridge, Canada)

Rodrigo Fernandes de Mello (Instituto de Ciências Matemáticas e de Computação)

Stan Franklin (University of Memphis, USA)

Akihiro Fujii (Kogakuin University, Japan)

Luis F. Goes (Pontifical Catholic University of Minas Gerais, Brazil)

Sergio Gonzalez-Rojo (LAAS-CNRS, France)

Enguerran Grandchamp (University of the French West Indies, France)

Jose Herrero (Universitat Politecnica de Catalunya, Spain)

Alexey Lastovetsky (University College Dublin, Ireland)

Bernard Lecussan (Supaero, France)

Laurent Manyri (LAAS-CNRS, France)

Carlos A. Martins (Pontifical Catholic University of Minas Gerais, Brazil)

Felix Mora-Camino (LAAS-CNRS, France)

Juan Navarro (Universitat Politecnica de Catalunya, Spain)

Akira Nishida (University of Tokyo, Japan)

Soufiane Noureddine (University of Lethbridge, Canada)

Yoshio Oyanagi (University of Tokyo, Japan)

Christiane Pousa (Pontifical Catholic University of Minas Gerais, Brazil)

Luiz Ramos (Pontifical Catholic University of Minas Gerais, Brazil)

Thomas Rauber (University of Bayreuth, Germany)

Gudula Rünger (Chemnitz University of Technology, Germany)

Luciano José Senger (Universidade Estadual de Ponta Grossa)

John Twamley (University College Dublin, Ireland)

Laurence T. Yang (St. Francis Xavier University, Canada)

Sebastian Zambal (Supaero, France)

Preface

This volume contains the papers selected for presentation at the International Symposium on High Performance Computational Science and Engineering 2004 in conjunction with IFIP World Computer Congress, held in Toulouse, France, August 27, 2004.

Computational Science and Engineering is increasingly becoming an emerging and promising discipline in shaping future research and development activities in academia and industry ranging from engineering, science, finance, economics, arts and humanitarian fields. New challenges are in the fields of modeling of complex systems, sophisticated algorithms, advanced scientific and engineering computing and associated (multi-disciplinary) problem solving environments. Because the solution of large and complex problems must cope with tight timing schedules, the use of high performance computing including traditional supercomputing, scalable parallel and distributed computing, emerging cluster and grid computing, is inevitable.

This event brings together computer scientists and engineers, applied mathematicians, researchers in other applied fields, industrial professionals to present, discuss and exchange idea, results, work in progress and experience of research in the area of high performance computational techniques for science and engineering applications.

Based on the review reports, 12 papers were accepted for publication in this volume. Additionally, this volumne contains the keynote talk given at the symposium.

We would like to express our gratitude to the members of the program committee as well as to all reviewers for their work.

MICHAEL NG, ANDREI DONCESCU, LAURENCE T. YANG, TAU LENG

I

KEYNOTE TALK

EXPLOITING MULTIPLE LEVELS OF PARALLELISM IN SCIENTIFIC COMPUTING

Thomas Rauber
Computer Science Department
University Bayreuth, Germany
rauber@uni–bayreuth.de

Gudula Rünger
Computer Science Department
Chemnitz University of Technology, Germany
ruenger@informatik.tu–chemnitz.de

Abstract Parallelism is still one of the most prominent techniques to improve the performance of large application programs. Parallelism can be detected and exploited on several different levels, including instruction level parallelism, data parallelism, functional parallelism and loop parallelism. A suitable mixture of different levels of parallelism can often improve the performance significantly and the task of parallel programming is to find and code the corresponding programs,

We discuss the potential of using multiple levels of parallelism in applications from scientific computing and specifically consider the programming with hierarchically structured multiprocessor tasks. A multiprocessor task can be mapped on a group of processors and can be executed concurrently to other independent tasks. Internally, a multiprocessor task can consist of a hierarchical composition of smaller tasks or can incorporate any kind of data, thread, or SPMD parallelism. Such a programming model is suitable for applications with an inherent modular structure. Examples are environmental models combining atmospheric, surface water, and ground water models, or aircraft simulations combining models for fluid dynamics, structural mechanics, and surface heating. But also methods like specific ODE solvers or hierarchical matrix computations benefit from multiple levels of parallelism. Examples from both areas are discussed.

Keywords: Task parallelism, multiprocessor tasks, orthogonal processor groups, scientific computing.

1. Introduction

Applications from scientific computing often require a large amount of execution time due to large system sizes or a large number of iteration steps. Often the execution time can be significantly reduced by a parallel execution on a suitable parallel or distributed execution platform. Most platforms in use have a distributed address space, so that each processor can only access its local data directly. Popular execution platforms are cluster systems, cluster of SMPs (symmetric multiprocessors), or heterogeneous cluster employing processors with different characteristics or sub-interconnection networks with different communication characteristics. Using standardized message passing libraries like MPI [Snir et al., 1998] or PVM [Geist et al., 1996], portable programs can be written for these systems. A satisfactory speedup is often obtained by a data parallel execution that distributes the data structures among the processors and lets each processor perform the computations on its local elements.

Many applications from scientific computing use collective communication operations to distribute data to different processors or collect partial results from different processors. Examples are iterative methods which compute an iteration vector in each iteration step. In a parallel execution, each processor computes a part of the iteration vector and the parts are collected at the end of each iteration step to make the iteration vector available to all processors. The ease of programming with collective communication operations comes for the price that their execution time shows a logarithmic or linear dependence on the number of executing processors. Examples are given in Figure 1 for the execution time of an MPI_Bcast() and an MPI_Allgather() operation on 24 processors of the cluster system CLIC (Chemnitzer Linux Cluster). The figure shows that an execution on the global set of processors requires a much larger time than a concurrent execution on smaller subsets. The resulting execution time is influenced by the specific collective operation to be performed, the implementation in the specific library, and the performance of the interconnection network used. The increase of the execution time of the communication operations with the number of processors may cause scalability problems if a pure data parallel SPMD implementation is used for a large number of processors. There are several techniques to improve the scalability in this situation:

(a) Collective communication operations are replaced by single-transfer operations that are performed between a single sender and a single receiver. This can often be applied for domain decomposition methods where a data exchange is performed only between neighboring processors.

(b) Multiple levels of parallelism can be exploited. In particular, a mixed task and data parallel execution can be used if the application provides task parallelism in the form of program parts that are independent of each other and that can therefore be executed concurrently. Depending on the application, multiple levels of task parallelism may be available.

(c) Orthogonal structures of communication can be exploited. In particular, collective communication operations on the global set of processors can be reorganized such that the communication operations are performed in phases on subsets of the processors.

Each of the techniques requires a detailed analysis of the application and, starting from a data parallel realization, may require a significant amount of code restructuring and rewriting. Moreover, not all techniques are suitable for a specific application, so that the analysis of the application also has to determine which of the techniques is most promising.

In this paper, we give an overview how multiple levels of parallelism can be exploited. We identify different levels of parallelism and describe techniques and programming support to exploit them. Specifically, we discuss the programming with hierarchically structured multiprocessor tasks (M-tasks) [Rauber and Rünger, 2000; Rauber and Rünger, 2002]. Moreover we give a short overview how orthogonal structures of communication can be used and show that a combination of M-task parallelism and orthogonal communication can lead to efficient implementations with good scalability properties. As example we mainly consider solution methods for ordinary differential equations (ODEs). ODE solvers are considered to be difficult to parallelize, but some of the solution methods provide task parallelism in each time step. The rest of the paper is organized as follows. Section 2 describes multiple levels of parallelism in applications from scientific computing. Section 3 presents a programming approach for exploiting task parallelism. Section 4 gives example applications that can benefit from using task parallelism. Section 5 shows how orthogonal structures of communication can be used. Section 6 demonstrates this for example applications. Section 7 concludes the paper.

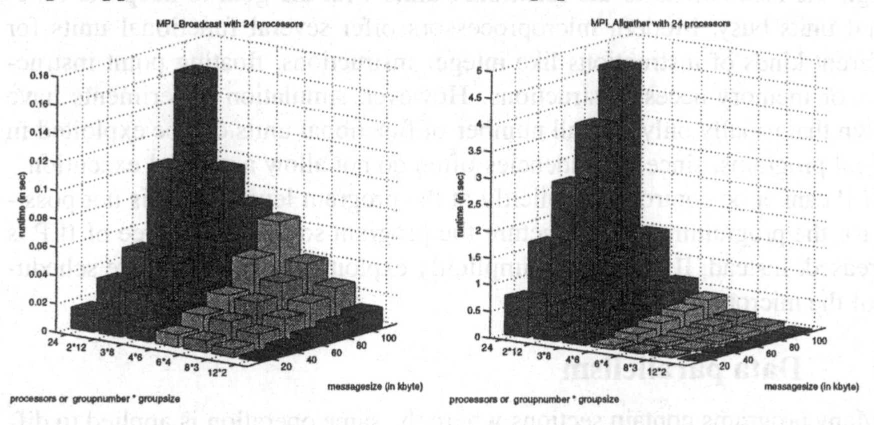

Figure 1. Execution time of MPI_Bcast() and MPI_Allgather() operations on the CLIC cluster. The diagrams show the execution time for different message sizes and groups organizations. The execution time for, e.g., group organization $2 * 12$ denotes the execution time on two groups of 12 processors, that work concurrently to each other.

2. Multiple levels of parallelism in numerical algorithms

Modular structures of cooperating subtasks often occur in large application programs and can be exploited for a task parallel execution. The tasks are often complete subprograms performing independent computations or simulations. The resulting task granularity is therefore quite coarse and the applications usually have only a small number of such subtasks. Numerical methods on the other hand sometimes provide potential task parallelism of medium granularity, but this task parallelism is usually limited. These algorithms can often be reformulated such that an additional task structure results. A reformulation may affect the numerical properties of the algorithms, so that a detailed analysis of the numerical properties is required. In this section, we describe different levels of parallelism in scientific applications and give an overview of programming techniques and support for exploiting the parallelism provided.

2.1 Instruction level parallelism

A sequential or parallel application offers the potential of instruction level parallelism (ILP), if there are no dependencies between adjacent instructions. In this case, the instructions can be executed by different functional units of the microprocessor concurrently. The dependencies that have to be avoided include true (flow) dependencies, anti dependencies and output dependencies [Hennessy and Patterson, 2003]. ILP results in fine-grained parallelism and is usually exploited by the instruction scheduler of superscalar processors. These schedulers perform a dependency analysis of the next instructions to be executed and assign the instructions to the functional units with the goal to keep the functional units busy. Modern microprocessors offer several functional units for different kinds of instructions like integer instructions, floating point instructions or memory access instructions. However, simulation experiments have shown that usually only a small number of functional units can be exploited in typical programs, since dependencies often do not allow a parallel execution.

ILP cannot be controlled explicitly at the program level, i.e., it is not possible for the programmer to restructure the program so that the degree of ILP is increased. Instead, ILP is always implicitly exploited by the hardware scheduler of the microprocessor.

2.2 Data parallelism

Many programs contain sections where the same operation is applied to different elements of large regular data structure like vectors or matrices. If there are no dependencies, these operations can be executed concurrently by different processors of a parallel or distributed system (data parallelism). Potential data parallelism is usually quite easy to identify and can be exploited by distributing the data structure among the processors and let each processor perform only the operation on its local elements (owner-computer rule). If the processors have to access elements that are stored on neighboring processors, a data exchange has to be performed before the computations to make these elements available. This is often organized by introducing ghost cells for each processor to store the elements sent by the neighboring processors. The data exchange has to be performed explicitly for platforms with a distributed address space by using a suitable message passing library like MPI or PVM, which requires an explicit restructuring of a (sequential) program.

Data parallelism can also be exploited by using a data parallel programming language like Fortran90 or HPF (High-Performance Fortran) [Forum, 1993] which use a single control flow and offer data parallel operations on portions of vectors or matrices. The communication operations to exchange data elements between neighboring processors do not need to be expressed explicitly, but are generated by the compiler according to the data dependencies.

2.3 Loop parallelism

The iterations of a loop can be executed in parallel if there are no dependencies between them. If all loop iterations are independent from each other, the loop is called a *parallel loop* and provides loop-level parallelism. This source of parallelism can be exploited by distributing the iterations of the parallel loop among the processors available. For a load balanced parallel execution, different loop distributions may be beneficial.

If all iterations of the loop require the same execution time, the distribution can easily be performed by a static distribution that assigns a fixed amount of iterations to each processor. If each iteration requires a different amount of execution time, such a static distribution can lead to load imbalances. Therefore dynamic techniques are used in this case. These techniques distribute the iterations in chunks of fixed or variable size to the different processors. The remaining iterations are often stored in a central queue from which a central manager distributes them to the processors. Non-adaptive techniques use chunks of fixed size or of a decreasing size that is determined in advance. Examples of non-adaptive techniques are FSC (fixed size chunking) and GSS (guided self scheduling) [Polychronopoulos and Kuck, 1987]. Adaptive techniques use chunks of variable size whose size is determined according to the number of remaining iterations. Adaptive techniques include factoring or weighted factoring [Banicescu and Velusamy, 2002; Banicescu et al., 2003] and allow also an adaptation of the chunk size to the speed of the executing processors. A good overview can be found in [Banicescu et al., 2003].

Loop-level parallelism can be exploited by using programming environments like OpenMP that provide the corresponding techniques or by implementing a loop manager that employs the specific scheduling technique to be used. Often, sequential loops can be transformed into parallel loops by applying loop transformation techniques like loop interchange or loop splitting, see [Wolfe, 1996] for an overview.

2.4 Task parallelism

A program exhibits task parallelism (also denoted as functional parallelism) if it contains different parts that are independent of each other and can therefore be executed concurrently. The program parts are usually denoted as tasks. Depending on the granularity of the independent program parts, the tasks can be executed as single-processor tasks (S-tasks) or multiprocessor tasks (M-tasks). M-tasks can be executed on an arbitrary number of processors in a data-parallel or SPMD style whereas each S-task is executed on a single processor.

A simple but efficient approach to distribute executable S-tasks among the processors is the use of (global or distributed) task pools. Tasks that are ready for execution are stored in the task pool from which they are accessed by idle processors for execution. Task pools have originally been designed for shared address spaces [Singh, 1993] and can provide good scalability also for irregular applications like the hierarchical radiosity method or volume rendering [Hoffmann et al., 2004]. To ensure this the task pools have to be organized in such a way that they achieve load balance of the processors and avoid bottlenecks when different processors try to retrieve executable tasks at the same time. Bottlenecks can usually be avoided by using distributed task pools that use a separate pool for each processor instead of one global pool that is accessed by all processors. When using distributed task pools, load balancing can be achieved by allowing processors to access the task pools of other processors if their local pool is empty (task stealing) or by employing a task manager that moves tasks between the task pools in the background. In both cases, each processor has to use synchronization also when accessing its local pool to avoid race conditions. The approach can be extended to distributed address spaces by including appropriate communication facilities [Hippold and Rünger, 2003].

The scheduling of M-tasks is more difficult than the scheduling of S-tasks, since each M-task can in principle be executed on an arbitrary number of processors. If there are several tasks that can be executed concurrently, the available processors should be partitioned into subsets such that there is one subset for each M-task and such that the execution of the concurrent M-tasks is finished at about the same time. This can be achieved if the size of the subsets of processors is adapted to the execution time of the M-tasks. The execution time is usually not known in advance and heuristics are applied.

M-tasks may also exhibit an internal structure with embedded M-tasks, i.e., the M-tasks may be hierarchically organized, which is a typical situation when executing divide-and-conquer methods in parallel.

3. Basics of M-task programming

This section gives a short overview of the Tlib library [Rauber and Rünger, 2002] that has been developed on top of MPI to support the programmer in the design of M-task programs. A Tlib program is an executable specification of the coordination and cooperation of the M-tasks in a program. M-tasks can be library functions or user-supplied functions, and they can also be built up from other M-tasks. Iterations and recursions of M-task specifications is possible; the parallel execution might result in a hierarchical splitting of the processor set until no further splitting is possible or reasonable. Using the library, the programmer can specify the M-tasks to be used by simply putting the operations to be performed in a function with a signature of the form

```
void *F (void * arg, MPI_Comm com, T_Descr *pdescr)
```

where the parameter **arg** comprises the arguments of the M-task, **comm** is the MPI communicator that can be used for the internal communication of the M-task and **pdescr** describes the current (hierarchical) group organization.

The Tlib library provides support for (a) the creation and administration of a dynamic hierarchy of processor groups, (b) the coordination and mapping of M-tasks to processor groups, (c) the handling and termination of recursive calls and group splittings and (d) the organization of communication between M-tasks. M-tasks can be hierarchically organized, i.e., each function of the form above can contain Tlib operations to split the group of executing processors or to assign new M-tasks to the newly created subgroups. The current group organization is stored in a group descriptor such that each processor of a group can access information about the group that it belongs to via this descriptor. The group descriptor for the global group of processors is generated by an initialization operation. Each splitting operation subdivides a given group into smaller subgroups according to the specification of the programmer. The resulting group structure is stored in the group descriptors of the participating processors. An example is the Tlib operation

```
int T_SplitGrp (T_Descr *pdescr, T_Descr *pdescr1,
                float p1, float p2)
```

with $0 \leq p1 + p2 \leq 1$. The operation generates two subgroups with a fraction **p1** or **p2** of the number of processors in the given processor group described by **pdescr**. The resulting group structure is described by group descriptor pdescr1. The corresponding communicator of a subgroup can be obtained by the Tlib operation

```
MPI_Comm T_GetComm (T_Descr *pdescr1).
```

Each processor obtains the communicator of the subgroup that it belongs to. Thus, group-internal communication can be performed by using this communicator. The Tlib group descriptor contains much more information including the current hierarchical group structure and explicit information about the group sizes, group members, or sibling groups.

The execution of M-tasks is initiated by assigning M-tasks to processor groups for execution. For two concurrent processor groups that have been created by T_SplitGrp(), this can be achieved by the Tlib operation

```
int T_Par (void * (*f1) (void *, MPI_Comm, T_Descr *),
           void * parg1, void * pres1,
           void * (*f2) (void *, MPI_Comm, T_Descr *),
           void * parg2, void * pres2,
           T_Descr *pdescr1).
```

Here, f1 and f2 are the M-task functions to be executed, parg1 and parg2 are their arguments and pres1 and pres2 are their possible results. The last argument pdescr1 is a group descriptor that has been returned by a preceeding call of T_SplitGrp(). The activation of an M-tasks by T_Par() automatically assigns the MPI communicator of the group descriptor, provided as last argument, to the second argument of the M-task. Thus, each M-task can use this communicator for group-internal MPI communication. The group descriptor itself is automatically assigned to the third argument of the M-task. By using this group descriptor, the M-task can further split the processor group and can assign M-tasks to the newly created subgroups in the same way. This allows the nesting of M-tasks, e.g. for the realization of divide-and-conquer methods or hierarchical algorithms.

In addition to the local group communicator, an M-task can also use other communicators to perform MPI communication operations. For example, an M-task can access the communicator of the parent group via the Tlib operations

```
parent_descr = T_GetParent (pdescr);
parent_comm = T_GetComm (Parent_descr);
```

and can then use this communicator to perform communication with M-tasks that are executed concurrently.

4. Examples for M-task programming

In this section, we describe some example applications that can benefit from an exploitation of task parallelism. In particular, we consider extrapolation methods for solving ordinary differential equations (ODEs) and the Strassen method for matrix multiplication.

4.1 Extrapolation methods

Extrapolation methods are explicit one-step solution methods for ODEs. In each time step, the methods compute r different approximations for the same point in time with different stepsizes $h_1, \ldots, h_r$ and combine the approximations at the end of the time step to a final approximation of higher order [Hairer et al., 1993; Deuflhard, 1985; van der Houwen and Sommeijer, 1990b]. The computations of the r different approximations are independent of each other and can therefore be computed concurrently as independent M-tasks. Figure 2 shows an illustration of the method.

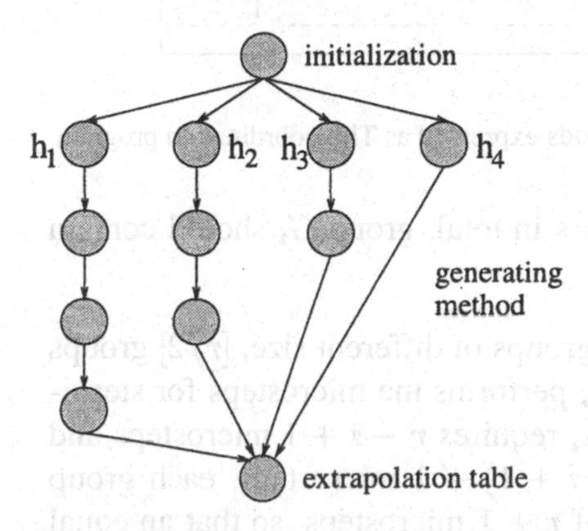

Figure 2. One time step of extrapolation method with $r = 4$ different stepsizes. The steps to perform one step with a stepsize h_i is denoted as microstep. We assume that stepsize h_i requires $r - i + 1$ microsteps of the generating method which is often a simple Euler method. The r approximations computed are combined in an extrapolation table to the final approximation for the time step.

The usage of different stepsizes corresponds to different numbers of microsteps, leading to a different amount of computation for each M-task. Thus, different group sizes should be used for an M-task version to guarantee that processor groups finish the computations of their approximations at about the same time. The following two group partitionings achieve good load balance:

(a) Linear partitioning: we use r disjoint processor groups $G_1, \ldots, G_r$ whose size $g_1, \ldots, g_r$ is determined according to the computational effort for the different approximations. If we assume that stepsize h_i requires $r - i + 1$ microsteps, a total number of $\sum_{i=1}^{r}(r-i+1) = (1/2)r \cdot (r+1)$ microsteps have to be performed. Processors group G_i has to perform i

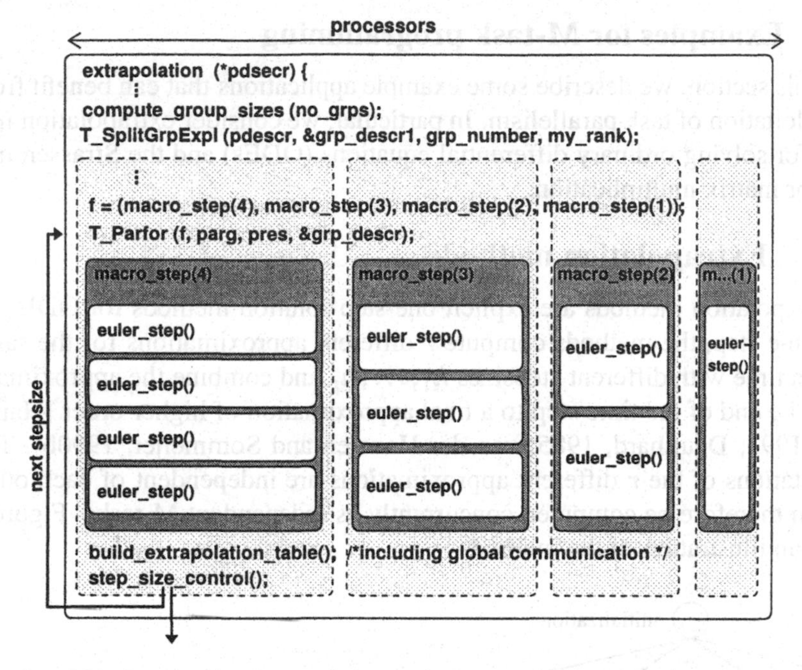

Figure 3. M-task structure of extrapolation methods expressed as Tlib coordination program.

steps of those. Thus, for p processors in total, group G_i should contain $g_i = p \cdot \frac{2 \cdot i}{r \cdot (r+1)}$ processors.

(b) Extended partitioning: instead of r groups of different size, $\lceil r/2 \rceil$ groups of the same size are used. Group G_i performs the microsteps for stepsizes h_i and h_{r-i+1}. Since stepsize h_i requires $r - i + 1$ microsteps and stepsize h_{r-i+1} requires $r - (r - i + 1) + 1$ microsteps, each group G_i has to perform a total number of $r + 1$ microsteps, so that an equal partitioning is adequate.

Figure 3 illustrates the group partitioning and the M-task assignment for the linear partitioning using the Tlib library. The groups are built using the Tlib function `T_SplitGrpExpl()` which uses explicit processor identifications and group identifications provided by the user-defined function compute_group_sizes(). The group partitioning is performed only once and is then re-used in each time step. The library call `T_Parfor()` is a generalization of `T_Par()` and assigns the M-tasks with the corresponding argument vectors to the subgroups for execution.

Figure 4 compares the execution time of the two task parallel execution schemes with the execution time of a pure data parallel program version on a Cray T3E with up to 32 processors. As application, a Brusselator ODE has been used. The figure shows that the exploitation of task parallelism is worth the

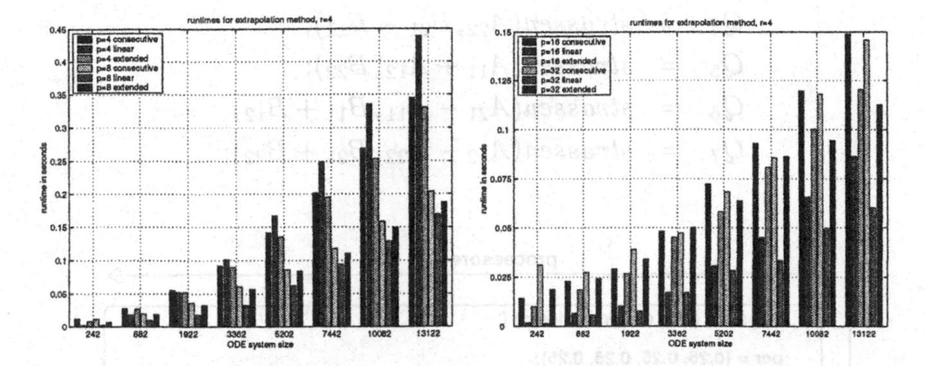

Figure 4. Execution time of an extrapolation method on a Cray T3E.

effort in particular for a large number of processors. When comparing the two task parallel execution schemes, it can be seen that exploiting the full extent of task parallelism (linear partitioning) leads to the shortest execution times for more than 16 processors.

4.2 Strassen matrix multiplication

Task parallelism can often be exploited in divide-and-conquer methods. An example is the Strassen algorithm for the multiplication of two matrices A, B. The algorithm decomposes the matrices A, B into square blocks of size $n/2$:

$$\begin{pmatrix} C_{11} & C_{12} \\ C_{21} & C_{22} \end{pmatrix} = \begin{pmatrix} A_{11} & A_{12} \\ A_{21} & A_{22} \end{pmatrix} \begin{pmatrix} B_{11} & B_{12} \\ B_{21} & B_{22} \end{pmatrix}$$

and computes the submatrices $C_{11}, C_{12}, C_{21}, C_{22}$ separately according to

$$\begin{aligned} C_{11} &= Q_1 + Q_4 - Q_5 + Q_7 \\ C_{12} &= Q_3 + Q_5 \\ C_{21} &= Q_2 + Q_4 \\ C_{22} &= Q_1 + Q_3 - Q_2 + Q_6 \end{aligned}$$

where the computation of the matrices $Q_1, \ldots, Q_7$ require 7 matrix multiplications for smaller matrices of size $n/2 \times n/2$. The seven matrix products can be computed by a conventional matrix multiplication or by a recursive application of the same procedure resulting in a divide-and-conquer algorithm with the following subproblems:

$$\begin{aligned} Q_1 &= strassen(A_{11} + A_{22}, B_{11} + B_{22}); \\ Q_2 &= strassen(A_{21} + A_{22}, B_{11}); \\ Q_3 &= strassen(A_{11}, B_{12} - B_{22}); \end{aligned}$$

$$\begin{aligned}Q_4 &= strassen(A_{22}, B_{21} - B_{11});\\ Q_5 &= strassen(A_{11} + A_{12}, B_{22});\\ Q_6 &= strassen(A_{21} - A_{11}, B_{11} + B_{12});\\ Q_7 &= strassen(A_{12} - A_{22}, B_{21} + B_{22});\end{aligned}$$

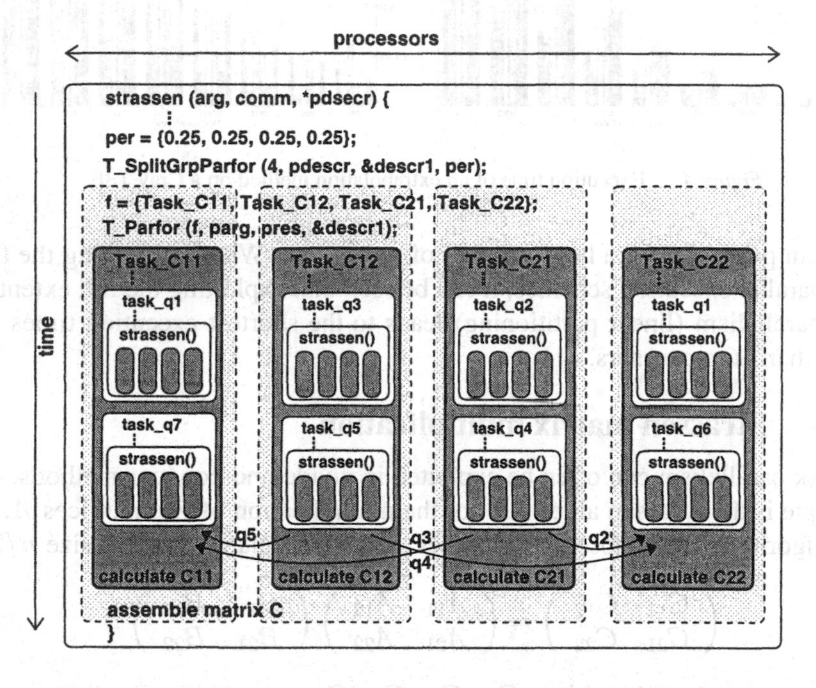

Figure 5. Task structure of the Strassen method using the Tlib library.

Figure 5 shows the structure of a task parallel execution of the Strassen algorithm and sketches a Tlib program. Four processor groups of the same size are formed by `T_SplitGrpParfor()` and `T_Parfor()` assigns tasks to compute C_{11}, C_{12}, C_{21} and C_{22}, respectively, to those groups. Internally, those M-tasks call other M-tasks to perform the subcomputations of $Q_1, \ldots, Q_7$, including recursive calls to Strassen. Communication is required between those groups and is indicated by annotated arrows. Such a task parallel implementation can be used as a starting point for the development of efficient parallel algorithms for matrix multiplication. In [Hunold et al., 2004b] we have shown that a combination of a task parallel implementation of the Strassen method with an efficient basic matrix multiplication algorithm tpMM [Hunold et al., 2004a] can lead to very efficient implementations, if a fast sequential algorithm like Atlas [Whaley and Dongarra, 1997] is used for performing the computations on a single processor.

5. Exploiting orthogonal structures of communication

For many applications, the communication overhead can also be reduced by exploiting orthogonal structures of communication. We consider the combination of orthogonal communication structures with M-task parallel programming and demonstrate the effect for iterated Runge-Kutta methods.

5.1 Iterated Runge-Kutta methods

Iterated Runge-Kutta methods (RK methods) are explicit one-step solution methods for solving ODEs, which have been designed to provide an additional level of parallelism in each time step [van der Houwen and Sommeijer, 1990a]. In contrast to embedded RK methods, the stage vector computations in each time step are independent of each other and can be computed by independent groups of processors. Each stage vector is computed by a separate fixed point iteration. For s stage vectors $v_1, \dots, v_s$ and right-hand side function f, the new approximation vector y_{k+1} is computed from the previous approximation vector y_k by:

$$
\begin{aligned}
v_{(0)}^l &= f(y_\kappa), \qquad l = 1, \dots, s \\
v_{(j)}^l &= f(y_\kappa + h_\kappa \sum_{i=1}^{s} a_{li} v_{(j-1)}^i), \quad l = 1, \dots, s, \quad j = 1, \dots, m \\
y_{\kappa+1}^{(m)} &= y_\kappa + h_\kappa \sum_{l=1}^{s} b_l \, v_{(m)}^l \\
y_{\kappa+1}^{(m-1)} &= y_\kappa + h_\kappa \sum_{l=1}^{s} b_l \, v_{(m-1)}^l
\end{aligned}
$$

After m steps of the fixed point iteration, $y_{(k+1)}^{(m)}$ is used as approximation for y_{k+1} and $y_{k+1}^{(m-1)}$ is used for error control and stepsize selection.

A task parallel computation uses s processor groups $G_1, \dots, G_s$ with sizes $g_1, \dots g_s$ for computing the stage vectors $v_1, \dots, v_s$. Since each stage vector requires the same amount of computations, the groups should have about equal size. An illustration of the task parallel execution is shown in Figure 5.1. For computing stage vector v_l, each processor of G_l computes a block of elements of the argument vector $\mu(l, j) = y_\kappa + h_\kappa \sum_{i=1}^{s} a_{li} v_{(j-1)}^i$ (where j is the current iteration) by calling `compute_arguments()`. Before applying the function f to $\mu(l, j)$ in `compute_fct()` to obtain $v_{(j)}^l$, each component of $\mu(l, j)$ has to be made available to all processors of G_l, since f may access all components of its argument. This can be done in `group_broadcast()` by a group-internal MPI_Allgather() operation. Using an M-task for the stage vector computation,

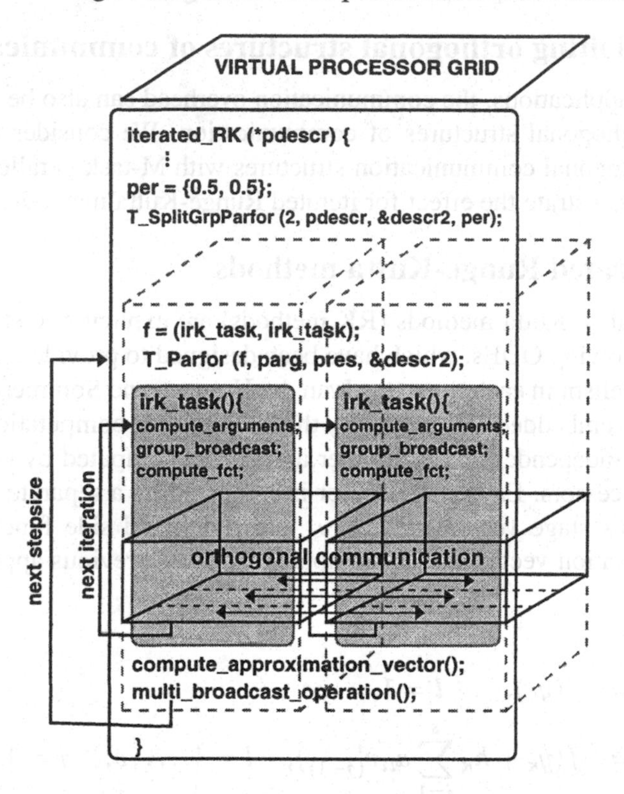

Figure 6. Exploiting orthogonal communication structures for iterated RK methods.

an internal communication operation is performed. Before the next iteration step j, each processor needs those parts of the previous iteration vectors $v^i_{(j-1)}$ to compute its part of the next argument vector $\mu(l, j)$. In the general case where each processor may have stored blocks of different size, the easiest way to obtain this is to make each processor the entire vectors $v^i_{(j-1)}$ available. This can be achieved by a global MPI_Allgather() operation. After a fixed number of iteration steps, the next approximation vector $y_{\kappa+1} = y^{(m)}_{\kappa+1}$ is computed and is made available to all processors by a global MPI_Allgather() operation.

5.2 Orthogonal Communication for Iterated RK Methods

For the special case that all groups have the same size g and that each processor stores blocks of the iteration vectors of the same size, orthogonal structures can be exploited for a more efficient communication. The orthogonal communication is based on the definition of orthogonal processor groups: For groups $G_1, \ldots, G_s$ with $G_l = \{q_{l1}, \ldots, q_{lg}\}$, we define orthogonal groups $Q_1, \cdots, Q_g$ with $Q_k = \{q_{lk} \in G_l, l = 1, \ldots, s\}$, see Figure 7. Instead of making all components of $v^i_{(j-1)}$ available to each processor, a group-based MPI_Allgather()

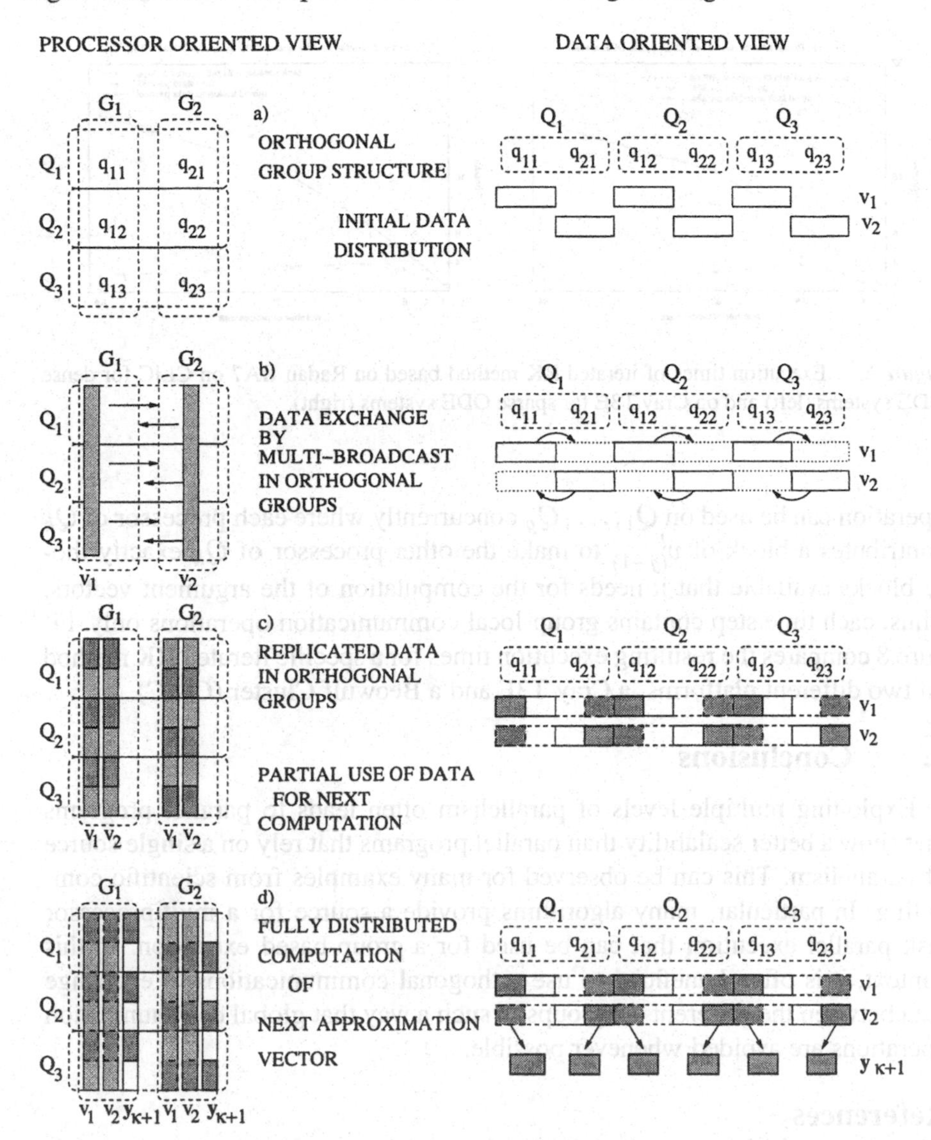

Figure 7. Exploiting orthogonal communication structures for the iterated RK method using two stage vectors v_1 and v_2 and corresponding processor groups $G_1 = \{q_{11}, q_{12}, q_{13}\}$ and $G_2 = \{q_{21}, q_{22}, q_{23}\}$. Part (a) illustrates the group structure with orthogonal groups Q_1, Q_2, Q_3 (left) and the distribution of v_1 and v_2 among the processors of G_1 and G_2. Part (b) illustrates the data exchange within the orthogonal groups after the computation of the argument vectors, leading to a replication of the corresponding blocks of v_1 and v_2 in the orthogonal groups, see part (c). Part (d) illustrates the usage of the stage vector blocks for the computation of the next approximation vector $y_{\kappa+1}$. Each processor uses only a part of the replicated data blocks.

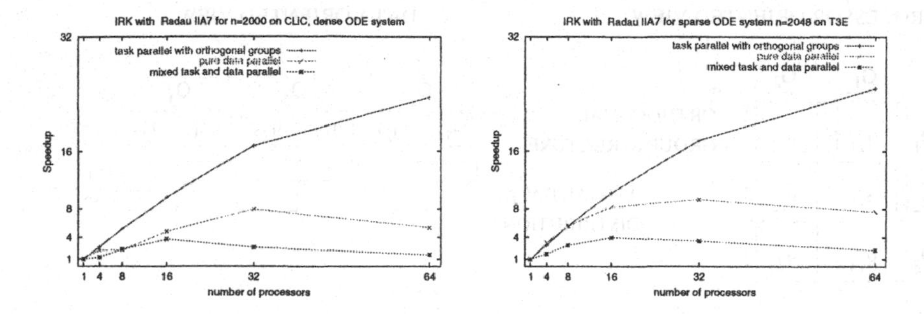

Figure 8. Execution times of iterated RK method based on Radau IIA7 on CLiC for dense ODE systems (left) and on Cray T3E for sparse ODE systems (right).

operation can be used on $Q_1, \ldots, Q_g$ concurrently where each processor of Q_k contributes a block of $v^l_{(j-1)}$ to make the other processor of Q_k exactly those blocks available that it needs for the computation of the argument vectors. Thus, each time step contains group-local communication operations only. Figure 8 compares the resulting execution times for a specific iterated RK method on two different platforms, a Cray T3E and a Beowulf Cluster (CLIC).

6. Conclusions

Exploiting multiple levels of parallelism often leads to parallel programs that show a better scalability than parallel programs that rely on a single source of parallelism. This can be observed for many examples from scientific computing. In particular, many algorithms provide a source for a multiprocessor task parallel execution that can be used for a group-based execution. In this context, it is often beneficial to use orthogonal communications to exchange data between the different subgroups in such a way that global communication operations are avoided whenever possible.

References

Banicescu, I. and Velusamy, V. (2002). Load balancing highly irregular computations with the adaptive factoring. In *Proc. of the IEEE - International Parallel and Distributed Processing Symposium (IPDPS 2002) - Heterogeneous Computing Workshop*. IEEE Computer Society Press, Fort Lauderdale.

Banicescu, I., Velusamy, V., and Devaprasad, J. (2003). On the Scalability of Dynamic Scheduling Scientific Applications with Adaptive Weighted Factoring. *Cluster Computing, The Journal of Networks, Software Tools and Applications*, 6(3):215–226.

Deuflhard, P. (1985). Recent progress in extrapolation methods for ordinary differential equations. *SIAM Review*, 27:505–535.

Forum, H. P. F. (1993). High Performance Fortran Language Specification. *Scientific Programming*, 2(1).

Geist, A., Beguelin, A., Dongarra, J., Jiang, W., Manchek, R., and Sunderam, V. (1996). *PVM Parallel Virtual Machine: A User's Guide and Tutorial for Networked Parallel Computing*. MIT Press, Cambridge, MA.

Hairer, E., Norsett, S., and Wanner, G. (1993). *Solving Ordinary Differential Equations I: Nonstiff Problems*. Springer–Verlag, Berlin.

Hennessy, J. and Patterson, D. (2003). *Computer Architecture — A Quantitative Approach*. Morgan Kaufmann, 3nd edition.

Hippold, J. and Rünger, G. (2003). Task Pool Teams for Implementing Irregular Algorithms on Clusters of SMPs. In *Proc. of the IPDPS (International Parallel and Distributed Processing Symposium)*, Nice, France. IEEE.

Hoffmann, R., Korch, M., and Rauber, T. (2004). Using Hardware Operations to Reduce the Synchronization Overhead of Task Pools. In *Proc. of the Int. Conference on Parallel Processing (ICPP)*, pages 241–249.

Hunold, S., Rauber, T., and Rünger, G. (2004a). Hierarchical Matrix-Matrix Multiplication based on Multiprocessor Tasks. In Bubak, M., van Albada, G., Sloot, P. M., and Dongarra, J. J., editors, *Proc. of the International Conference on Computational Science ICCS 2004, Part II*, LNCS 3037, pages 1–8. Springer.

Hunold, S., Rauber, T., and Rünger, G. (2004b). Multilevel Hierarchical Matrix Multiplication on Clusters. In *Proc. of the 18th Annual ACM International Conference on Supercomputing, ICS'04*, pages 136–145.

Polychronopoulos, C. and Kuck, D. (1987). Guided self-scheduling: A practical scheduling scheme for parallel supercomputers. *IEEE Transactions on Computers*, C-36(12):1425–1439.

Rauber, T. and Rünger, G. (2000). A Transformation Approach to Derive Efficient Parallel Implementations. *IEEE Transactions on Software Engineering*, 26(4):315–339.

Rauber, T. and Rünger, G. (2002). Library Support for Hierarchical Multi-Processor Tasks. In *Proc. of the Supercomputing 2002*, Baltimore, USA. ACM/IEEE.

Singh, J. (1993). *Parallel Hierarchical N-Body Methods and their Implication for Multiprocessors*. PhD thesis, Stanford University.

Snir, M., Otto, S., Huss-Ledermann, S., Walker, D., and Dongarra, J. (1998). *MPI: The Complete Reference, Vol.1: The MPI Core*. MIT Press, Camdridge, MA.

van der Houwen, P. and Sommeijer, B. (1990a). Parallel Iteration of high–order Runge–Kutta Methods with stepsize control. *Journal of Computational and Applied Mathematics*, 29:111–127.

van der Houwen, P. and Sommeijer, B. (1990b). Parallel ODE Solvers. In *Proc. of the ACM Int. Conf. on Supercomputing*, pages 71–81.

Whaley, R. C. and Dongarra, J. J. (1997). Automatically Tuned Linear Algebra Software. Technical Report UT-CS-97-366, University of Tennessee.

Wolfe, M. (1996). *High Performance Compilers for Parallel Computing*. Addison Wesley.

II

DISTRIBUTED COMPUTING

A Sparse Distributed Memory Capable of Handling Small Cues, SDMSCue

Ashraf Anwar and Stan Franklin
College of Computing and Information Technology, Arab Academy for Science, Technology and Maritime Transport, Alexandria, Egypt, dr.a.anwar@ccit.aast.edu; Computer Science Department, University of Memphis, Memphis, TN 38152, franklin@memphis.edu

Abstract: In this work, we present Sparse Distributed Memory for Small Cues (SDMSCue), a new variant of Sparse Distributed Memory (SDM) that is capable of handling small cues. SDM is a content-addressable memory technique that relies on similar memory items tending to be clustered together in the same region or subspace of the semantic space. SDM has been used before as associative memory or control structure for software agents. In this context, small cues refer to input cues that are presented to SDM for reading associations; but have many missing parts or fields from them. The original SDM failed to handle such a problem. Hence, our work with SDMSCue comes to overcome this pitfall. The main idea in our work; is the projection of the semantic space on a smaller subspace; that is selected based on the input cue pattern, to allow for read/write using an input cue that is missing a large portion. The test results show that SDMSCue is capable of recovering and recalling information from memory using an arbitrary small part of that information; when the original SDM would fail. SDMSCue is augmented with the use of genetic algorithms for memory allocation and initialization. We think that the introduction of SDMSCue opens the door to more research areas and practical uses for SDM in general.

Keywords: Artificial Intelligence; Cognition; Memory; SDM; SDMSCue; Software Agents.

1. INTRODUCTION

Sparse Distributed Memory (SDM) is a content addressable memory developed by Kanerva. SDM was proposed to be a tool and model of human associative memory (Kanerva, 1988a; Kanerva & Raugh, 1987).

SDM has proven successful in modeling associative memories (Anwar, 1997; Anwar & Franklin, 2003; Rao & Fuentes, 1996; 1998; Scott, Fuller, & O'Brien, 1993). Associative memory is typically needed for intelligent and cognitive autonomous agents (Glenberg, 1997; Kosslyn, 1992). In particular, both cognitive software agents (Franklin, 1997; 2001) and "conscious" software agents (Franklin & Graesser, 1999) need such a memory. One of the "conscious" software agents, which we did work with, IDA: Intelligent Distribution Agent, uses SDMSCue (Franklin, Kelemen, & McCauley, 1998). The use of SDMSCue in IDA is to learn and keep associations between various pieces of information pertaining to the task of personnel distribution.

2. THE MOTIVE FOR SDMSCue

In many cases the need for associative memory to be able to handle and retrieve associations based on arbitrary small cues is crucial. For example; in IDA, we are often faced with a situation in which we need to retrieve associations based on very small pieces of information like part of email address, part of name, or social security number. Humans have no problem retrieving associations based on arbitrary small cues. While the original SDM modeled many aspects of human memory very successfully, it failed miserably in dealing with the issue of retrieving associations for short-length or small cues. Without such a capability, we are missing a key human-like feature in associative memory models that are based on SDM. Hence, the role of SDMSCue comes to the scene.

SDMSCue uses an elegant space projection mechanism to *enlarge* the short-length input cue successively until it is large enough for a read/write from/to the entire full-length SDM semantic space. The enlargement process uses successively increasing subspaces for reads/writes. To be noted is that both read and write operations in SDM involve the selection of an access circle to read from, or to write to. The selection is typically based on similarity between the input read/write cue, and the hard locations addresses within the access circle.

3. SPARSE DISTRIBUTED MEMORY

Sparse Distributed Memory, SDM, is the work of Pentti Kanerva (1988a, 1988b). It gets its name from the sparse allocation of storage locations in a vast binary address space and from the distributed nature of information storage and retrieval. A typical SDM has a vast binary space of possible memory locations in a 2^n semantic space where n is both the full word length and the dimension of the address space. For any practical application, only a very small portion of this 2^n semantic space can actually exist. For more details and discussions, see (Anwar, 1997). Also see (Franklin, 1995) for a brief overview, and (Willshaw, 1990) for a useful commentary. Many evaluations, extensions, and enhancements have been suggested for SDM (Evans & Surkan, 1991; Karlsson, 1995; Kristoferson, 1995a; 1995b; Rogers, 1988a; 1988b; Ryan & Andreae, 1995). A more efficient initialization technique for SDM using Genetic Algorithms was also suggested (Anwar, Dasgupta, & Franklin, 1999).

There are two main types of associative memory, auto-associative and hetero-associative. In auto-associative memory, a memory item, typically with noise and/or missing parts, is used to retrieve itself. In hetero-associative memory, memory items are stored in sequences where one item leads to the next item in the sequence. The auto-associative version of SDM is truly an associative memory technique where the contents and the addresses belong to the same space and are used alternatively.

A Boolean space is the set of all Boolean vectors (points) of some fixed length, n, called the dimension of the space. The Boolean space of dimension n contains 2^n Boolean vectors, each of length n. The number of points increases exponentially as the dimension increases. Boolean geometry uses a metric called *Hamming Distance,* where the distance between two points is the number of coordinates at which they differ. Thus d((1,0,0,1,0), (1,0,1,1,1)) = 2. The distance between two points will measure the similarity between two memory items, closer points being more similar. We may think of these Boolean vectors as feature vectors, where each feature can be only on, 1, or off, 0. Two such feature vectors are closer together if more of their features are the same.

The word length, which is also the dimension of the space, determines how rich in features each word and the overall semantic space are. Features are represented by one or more bits in a Boolean vector or binary string of length n. Groups of features are concatenated to form a word, which becomes a candidate for writing/reading into/from SDM. Another important factor in SDM design; is how many real memory locations are implemented. These are called *hard locations*. When writing, a copy of the object binary string is placed in all -close enough- hard locations. When reading, a subject cue would reach all close enough hard locations and get some sort of aggregate or average word from them. Reading is not always successful (Anwar, 1997; Kanerva, 1988a). Depending on the cue and the previously written information, among other factors, convergence or divergence during an iterative reading operation may occur. If convergence occurs, the pooled word will be the closest match of the input reading cue, possibly with some sort of abstraction. On the other hand, when divergence occurs, there is no relation, in general, between the input cue and what is retrieved from SDM.

SDM is a content addressable memory that, in many ways, is ideal for use as a long-term associative memory. "Content addressable" means that items in memory can be retrieved by using all or part of their contents as an address or a cue; rather than having to know their actual addresses in memory as in traditional computer memory.

Envisioning the semantic space as a sphere around an arbitrary point, and for n sufficiently large; most of the address space lies midway in the sphere from the point at the center of the sphere (Kanerva, 1988a). In other words, almost all the space is far away from any given point.

A Boolean space implementation is typically sparsely populated for sufficiently large n; an important property for the construction of SDM, and the source of part of its name. For n=1000, one cannot hope to actually implement such a vast 2^n memory in its entirety. On the other hand, considering humans and feature vectors, a thousand features wouldn't deal with just human visual input until a high level of abstraction has been reached. A dimension of 1000 may not be all that much for real-life cognition; it may, for some purposes, be unrealistically small. Kanerva proposes to deal with this vast address space by choosing a uniform random sample, of size 2^{20} for n = 1000, of hard locations. An even better way to distribute the set of hard locations over the vast semantic space using Genetic Algorithms was suggested (Anwar, 1999). With 2^{20} hard locations out of 2^{1000} possible semantic locations, the ratio is 2^{-980}, *truly sparse*.

SDM is distributed in that many hard locations participate in storing and retrieving (write/read) each word, and one hard location can be involved in the storage and retrieval of many words. This is very different from the one-word-per-location type of memory to which we are accustomed. For n = 1000, each hard location, stores data in 1000 counters, each with range from –K to K, where K is an implementation-dependent factor that determines memory capacity. We now have about a million hard locations, each with a thousand counters, totaling a billion counters in all. For K = 40, numbers in the range -40 to 40 will take most of a byte to store. Thus we are talking about a billion bytes, a gigabyte, of memory.

Counters are updated according to the words written. Writing 1 to one counter increments it; writing 0 decrements it. To write a word at a given hard location x, write each coordinate of the word into the corresponding counter in x; either incrementing it or decrementing it.

At any arbitrary location, the sphere centered at that location, is called the access sphere or access circle of that location. For n = 1000, and 2^{20} hard locations, an access sphere typically contains about a thousand hard locations, with the closest usually some 424 bits away and the median distance from the center point to hard locations in the access sphere about 448. Any hard location in the access sphere is said to be *accessible* from the center point address or word. With this machinery in hand, any location -hard or not- can be written to in a distributive manner. To write a word to a location, simply write the word to each of the roughly one thousand hard locations within the location access sphere, i.e. accessible from the location as a center point.

We read from an arbitrary point or address in the semantic space. This read includes reading from all hard locations within the access sphere from that point. To read from a single hard location, x, we compute the bit vector read at x, by assigning its ith bit the value 1 or 0 according as the ith counter at x is positive or negative. Thus, each bit results from a majority rule decision of all the data that have been written at x. The read word is an archetype of the words that have been written to x, but may not be any one of them. For any location or address, the bit vector read is formed by pooling the data read from each hard location accessible from that location or address. Each bit of the read word results from a majority rule decision over the pooled data. The voting is influenced by n stored copies of the word, and about (10 n) other stored data items. Since the intersection of two access spheres is typically quite small, the other data items influence a given coordinate only in small groups of ones or zeros, which tend to compensate for each other, i.e. white noise. The n copies of the original word drown out this slight noise.

The entire stored word is not needed to recover itself. Iterative reading allows recovery when reading from a noisy version of what has been stored. There are conditions, involving how much of the stored word is available for the read operation; under which this is true, (Kanerva, 1988a). Reading with a cue that has never been written to SDM before gives, if convergent, the closest match stored in SDM to that input cue, with some sort of abstraction if close items have been written to the memory. SDM works well for reconstructing individual memories (Hely, 1994).

4. SDM FOR SMALL CUES (SDMSCue)

4.1 Approach

Using a variant of SDM capable of handling small cues, we are able to overcome the main shortage in Kanerva's model (Kanerva, 1988a; 1992). One of the main problems with Kanerva's SDM is that the input cue has to be of sufficient length to be able to retrieve a match. The reason is that the *entire* input cue is considered and the hamming distance between its *entire* binary string representation and various hard locations in the access sphere or access circle is considered when reading or writing. So if we have a small cue, the missing large part is almost guaranteed to sink the known small cue in terms of hamming distance, thus being indifferent to all words or hard locations in the SDM memory.

In many cases, we are faced with a very distinguished and unique memory cue that is considerably small in size than the typical 65-80% requirement in Kanerva's work. We -humans- are able to retrieve relevant information associated with such a small cue efficiently. For SDM to be able to function similarly, we need a variant of SDM that is capable of handling small cues. When faced with retrieval based on a substantially smaller cue like part of a name, part of an email address, or SSN, this calls for the use of SDMSCue.

The goal is to retrieve appropriate corresponding word matching such a small cue. Using subspaces with increasing sizes in a progressive way, we are able to read and retrieve the whole original corresponding memory item using only a small portion of the cue with arbitrary small length. However, the number of levels needed for read/write depends upon the original size of the small cue. The smaller the original input cue is; the more levels of read/write needed. Time complexity of the operation is linearly proportional to the number of levels needed.

4.2 Design of SDMSCue

The idea is to project the original SDM semantic space onto a smaller subspace corresponding to the small cue. In such a projection, only memory locations with ***matching*** content to the small input cue contribute to Read/Write operations.

By projecting the space onto a smaller subspace, we are able to use the smaller subspace for much higher recall rate for a considerably small cue. The gain occurs mainly because of constraining and limiting only hard locations that match the small cue part and allowing only them to contribute to the subspace for read/write operations.

The result obtained from a read/write operation at one stage; is used to access a larger subspace including the input cue along with associations retrieved that typically range from 25% to 35% of the former input cue length. Such associations are retrieved from the contents of the hard locations that were selected, and contributed to the former subspace read/write.

By repeating the above process for increasingly larger subspaces and levels of projection, we eventually get to access the entire semantic space for read/write.

To be noted is that during write operations, actual writing to hard locations occurs only at the final level when writing to the entire space. All preceding access takes place for association retrieval only. So both read and write operations are the same (reading and retrieving associations) until the final level when we access the entire space. In both cases association buildup takes place to enlarge the small input cue gradually until the length obtained is large enough to read from the entire semantic space. In the last level or phase, if it is a read operation, we simply retrieve associations and obtain the matching entire word. If it is a write operation, the enlarged input cue is written to all hard locations within the access circle of the last phase, which is part of the entire semantic space.

4.3 How SDMSCue Works

Using SDMSCue, we can manage to access (read/write) with small cues. The process goes in phases in reading or writing operations. When accessing, in the first phase, we read from a small sub-space that corresponds to the input small cue plus extra association information. This read –if convergent- yields a longer word due to the association of information. This resulting word is then used as the input to the second phase. In the second phase, a similar process takes place reading from a larger subspace using the output result from the first phase as input. This process continues until the subspace being read from is the entire original semantic space.

For example, as shown in *Table 1*, we start by reading with a small cue of length 17% of the whole memory word size, using a 0.35 ratio for associations. Then the reading operation yields a larger word, due to adding associations, of length 23% of the whole memory word.

*Table 1: Multi-Level Reading operations from SDMSCue, and their corresponding Length Percentage, **m**. In Level 1, with a small cue of length 17% of the whole memory word, projection of the space and reading yields bigger subspace of 23%. In a similar way successive projection increases the subspace till it reaches the whole space, i.e. 100% of the memory word.*

Level #	Input Word Length	Output Word Length	Output Word
1	17	23	|---------------|
2	23	31	|---------------------|
3	31	42	|----------------------------|
4	42	57	|--------------------------------------|
5	57	77	|---|
6	77	100	|--|

Then using the 23% retrieved and formed word as input to the second phase and adding 0.35 associations to it, a 31% word is obtained. This process continues until in the final level (6th level in the example), a 77%

retrieved and formed word is used to access (read from or write to) the entire original semantic space.

To be noted is that the time complexity of a Read/Write operation is ***linearly*** proportional to the number of levels involved in a Read/Write. However, the overall effect is quite minor compared to the gain of the approach. Assume an original cue length of ***m%*** of the whole memory word size; where **m** ranges from 0 to 100. When reading/writing from SDMSCue, some associations are retrieved for the small cue at each level resulting in a length gain. Let ***i*** be the percentage of the length gain at each level. Adding the gain in length, ***i***, to the next input cue in each successive read/write level, the maximum number of read/write levels ***N*** is given by:

$100 \leq m * (1 + i)^N$; Last word length needs to be 100%, i.e. last read needs to occur from the entire semantic space

$$\Rightarrow N = \lceil \log(100/m)/\log(1+i) \rceil$$
$$\Rightarrow N = \lceil (2 - \log m)/\log(1+i) \rceil$$

For $m = 17$ (17% original small cue length), $i = 0.35$ (35%), as in *Table 1*, $N = \lceil (2 - \log 17) / \log 1.35 \rceil = \lceil 5.9 \rceil = 6$ Levels.

For $m = 10$ (10% original small cue length), $i = 0.3$ (30%), $N = \lceil (2 - \log 10) / \log 1.3 \rceil = \lceil 8.78 \rceil = 9$ Levels.

For $m = 1$ (1% original small cue length), $i = 0.3$ (30%), $N = \lceil (2 - \log 1) / \log 1.3 \rceil = \lceil 17.55 \rceil = 18$ Levels.

We define the term SDMSCue **Latency Factor** to be the average number of levels needed for read/write for a certain word distribution to be written to or read from SDMSCue. Such a factor is both semantic space dependent, and distribution dependent.

SDMSCue makes use of GA for more efficient space initialization and hard locations allocation (Anwar, 1999; Anwar, Dasgupta, & Franklin, 1999). The uniformity of the semantic space is –in general- favorable to better recall rates for SDMSCue as well as SDM.

4.4 SDMSCue Convergence and Divergence

We need to develop a notion for overall convergence and divergence in case of Read/Write from SDMSCue. For overall convergence to occur, all phases or levels of Read/Write must converge. Overall divergence occurs if at any phase or level, the Read/Write diverges. In other words, convergence is the Boolean "AND" operation of the convergence in all levels.

$$\text{Convergence}\,|_{\text{SDMSCue}} = \text{AND}_{\text{For All } i}(\text{convergence}\,|_{\text{at level } i})$$

$$\text{Divergence}\,|_{\text{SDMSCue}} = \text{NOT}\,(\text{Convergence}\,|_{\text{SDMSCue}})$$

4.5 Implementation

The following is a short note about the current implementation of SDM with small cues (SDMSCue). Java Visual Symantec Caf Professional Edition was used for testing and implementation of the code for SDMSCue in Windows XP environment. The hardware was a Pentium 2.4 GHz with 1GB RAM. The results obtained are based on recall performance and memory trace used for comparison tests of SDMSCue vs. original SDM. Runs were performed repetitively 100 times on average for each case.

5. RESULTS AND STATISTICS

The following comparison between SDMSCue and regular SDM was done using the same memory parameters, and memory trace (Loftus & Loftus, 1976). Memory performance in terms of various operational parameters was considered for SDMSCue vs. original SDM. The various memory parameters: Memory Volume, Cue Volume, Similarity, and Noise were considered.

Memory Volume is the average number of features in the memory trace. In other words, it is the average number of 1's in a memory word. It measures the richness of the memory trace. Memory volume is a vital parameter in the distinction of the memory trace. It signifies the distribution of various memory words over the semantic space.

Retrieval Volume is the same as Memory volume but for a single input cue or input word to memory. It has almost the same effect on retrieval as memory volume.

Similarity is a measure of how similar, in average, are the words written to memory. The more similar the words written to memory are, the more clustered contiguously they are, and the harder it is to retrieve them. The hamming distance is the measure of similarity in SDM as well as SDMSCue. The less the hamming distance between two memory words, the more similar the memory words are. However, there is a difference between the similarity of hard locations and the similarity of written memory words. Using genetic algorithms (Anwar, 1999) a uniform distribution of the hard locations in SDM can be obtained.

Noise determines the number of noise bits, on average, in a memory word. It reflects directly on the reliability of retrieval of stored memory words.

Table 2 shows the distribution of the percentage of input cues in memory trace, used for the test, with respect to cue length, measured as percentage of the whole length. For example, according to *Table* 2, 35% of the cues in memory trace do not exceed 20% in length (small cues), while 25% of the cues in memory trace have length greater than 20% but less than or equal to 40% (low medium cues). Also only 10% of the input cues have length greater than 70% (longest cues). The second column gives the number of levels needed for read/write operation using the formula devised in section 4.3. As shown in *Table 2*, for the chosen distribution, the overall average cue length is 36%, and the average number of levels for read/write is 5. So, for the distribution at hand, SDMSCue has a latency factor of 5.

Table 2: The distribution of input cues in memory trace with respect to cue length, and their corresponding number of Read/Write levels.

Cue Minimum-Maximum Length as percentage of the Whole Word	Average Number of Read/Write Levels Needed	Percentage of Cues in Memory Trace
Less than 20%	8	35%
20%-40%	5	25%
40%-50%	3	10%
50%-60%	2	10%
60%-70%	2	10%
70%-100%	1	10%
Average Length 36%	**Latency Factor** = Average Number of Levels = **5**	

To be noted is that this distribution was chosen to illustrate the advantage of using SDMSCue when considerable percentage of the input cues is short in length, i.e. missing too many parts. By no means is this the only distribution that can illustrate the idea, but just the one we settled upon after some trials to illustrate the benefit of using SDMSCue when considerable number of the input cues is short in length. However, varying the distribution will definitely change the gain achieved from using SDMSCue over SDM.

Table 3 shows a comparison between the recall in SDM vs. SDMSCue. Various combinations of the memory trace parameters were considered. Each was varied on a Low/High scale.

Table 3: Comparison between GA initialized SDM, and GA initialized SDMSCue. This is a comparison between the performance of SDM with and without small Cues Capability, and the gain in memory recall resulting from the use of SDMSCue. Memory Volume and Retrieval Volume (H=60%, L=10%). Similarity between Memory Items (H=70%, L=30%). Noise (H=30%, L=10%).

#	Memory Volume	Retrieval Volume	Similarity	Noise	SDM Hit % in Recall	SDMSCue Hit % in Recall	Recall Gain %	Decrease in Miss in Recall Gain %
1	L	L	L	L	6	44	633	40
2	L	L	L	H	5	40	700	37
3	L	L	H	L	7	39	457	34
4	L	L	H	H	5	34	580	31
5	L	H	L	L	9	56	522	52
6	L	H	L	H	8	50	525	46
7	L	H	H	L	11	49	345	43
8	L	H	H	H	5	42	740	39
9	H	L	L	L	74	98	32	92
10	H	L	L	H	66	93	41	79
11	H	L	H	L	61	93	52	82
12	H	L	H	H	54	89	65	76
13	H	H	L	L	73	99	36	96
14	H	H	L	H	56	95	70	89
15	H	H	H	L	60	96	60	90
16	H	H	H	H	50	92	84	84

The gain achieved from using SDMSCue is illustrated in the last two columns. The first gain, Recall Gain, measures the improvement in successful recall or Hit in SDMSCue over original SDM. The second gain, Decrease in Miss in Recall, measures the decrease in Miss in SDMSCue over the original SDM. This gain measures the improvement in SDMSCue over SDM in terms of decrease in the percentage of unsuccessful recall or memory Miss. To be noted is that a T-Test of statistical dependence (Kanji, 1999; Vogt, 1998) shows statistical significance between the recall of SDMSCue and that of SDM.

In *Table* 3, Row 4, which represents poor recall conditions (Low Memory Volume, Low Retrieval Volume, High Similarity, and High Noise), shows large improvement in successful recall in SDMSCue over original SDM. Row 13, on the other hand, represents near optimal recall conditions (High Memory Volume, High Retrieval Volume, Low Similarity, and Low Noise). Row 13 corresponds to 36% recall gain.

For decrease in miss in recall gain, recall conditions play similar role. Row 4, which represents poor recall conditions, shows 31% gain. Row 13, which represents near optimal recall conditions, shows 96% gain.

In general, results in *Table 3* show that the higher the volume of the memory and/or the retrieval volume, the better the recall for both memories, SDMSCue and SDM. However, when the memory volume is quite low, the recall gets really affected. The degradation in performance is more graceful in SDMSCue than it is in SDM. This has to do with the fact that SDMSCue projects the space on the part of the cue that exists, thus greatly moderating the typical negative effect of low memory volume.

With respect to similarity, the more distinct the memory words are, i.e. the less similarity, the better the recall in general. This makes absolute sense since SDM in general; and accordingly SDMSCue as well, uses hamming distance as a measure of inclusion of memory words in access sphere or access circle for read/write. The more similar memory words are, the closer they become in terms of hamming distance. Hence, they tend to get clustered together and cause crowding effect in the semantic space, where they may sink each other in certain regions of the semantic space.

To be noted, though, is that such clustering effect in the semantic space resulting from highly similar items being written to SDM or SDMSCue is somewhat similar but not the same as clustering of hard locations when a poor initialization technique is used for hard locations assignment. The later is more crucial to the functioning of SDM or SDMSCue, and should be application independent, whereas the former depends on the memory trace fed to SDM or SDMSCue, and hence is application dependent by nature. While there is a straightforward way to guarantee uniformity of hard locations assignment in the semantic space of SDM or SDMSCue (Anwar, Dasgupta, & Franklin, 1999), there is no immediate direct solution to overcome clustering in SDM or SDMSCue due to similarity of words or memory items written to it.

As expected, for the effect of noise on recall; the lower the noise, the better the recall in general. Noise can however be sunk to a degree as a result of the distributed nature of read/write, as well as the abstraction achieved from using SDM and SDMSCue.

Figure 1 shown below contrasts the performance of SDMSCue vs. SDM in terms of hit rate in recall. In this comparison, the 16 different memory configurations in *Table 3* were used, e.g. configuration 9 is HLLL which stands for High Memory Volume, Low Retrieval Volume, Low Similarity, and Low Noise. For each configuration, the same memory trace was applied to both SDM and SDMSCue. Then recall was tested with the same set of patterns with lengths distributed according to *Table 2*. Recall hit rate for each memory parameters configuration (recall conditions combination) was calculated and is shown as percentage over the vertical axis in *Figure 1*.

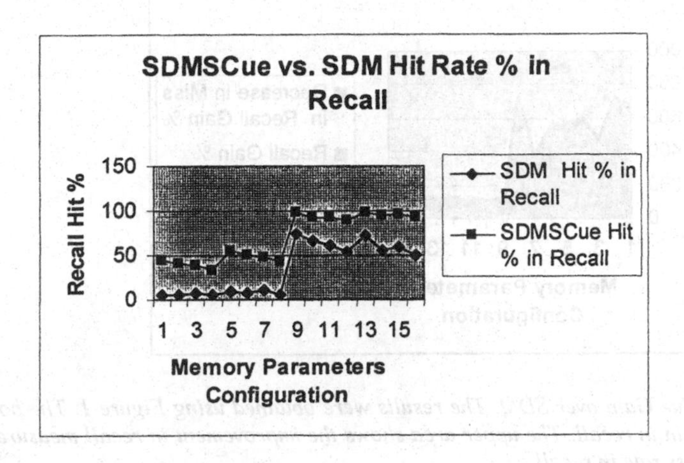

Figure 1: SDMSCue vs. SDM Hit Rate % in Recall. The top line shows the recall hit rate in SDMSCue. The bottom line shows the recall hit rate in SDM. The X-axis is for the 16 different memory parameters configurations from Table 3. The Y-axis is for the percentage of Recall Hit Rate.

It is clear from *Figure 1* that SDMSCue consistently outperforms SDM in terms of recall under all conditions. This comes at no surprise since SDMSCue uses SDM functionality in addition to the elegant space-projection to filter out non-relevant memory locations. SDMSCue also uses a far more superior GA approach for uniform space initialization and allocation of hard locations (Anwar, 1999).

Figure 2 shows the gain of SDMSCue over original SDM using the results and figures from *Figure 1*. The bottom area is the recall gain in SDMSCue over SDM. The gain percentage of SDMSCue over SDM is defined by: 100*(SDMSCue Hit% – SDM Hit%) / (SDM Hit%). For example, for memory parameter configuration 1, the gain is 100 * (44 - 6) / 6 = 633%.

The upper area in *Figure 2* is the improvement in recall measured as the decrease in miss rate in recall. This is defined by: 100*(SDM Miss% – SDMSCue Miss%) / (SDM Miss%). For example, for memory parameter configuration 1, the improvement as decrease in miss rate is 100*(94 – 56) / 94 = 40%. For memory parameter configuration 13, the improvement as decrease in miss rate is 100*(27 – 1) / 27 = 96%.

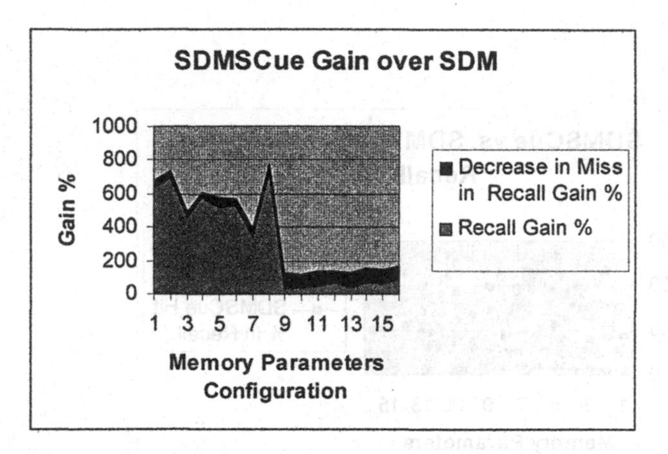

Figure 2: SDMSCue Gain over SDM. The results were obtained using Figure 1. The bottom area shows the gain in recall. The upper area shows the improvement in recall measured as the decrease in miss rate in recall.

6. CONCLUSIONS

SDMSCue is superior to traditional SDM in its capability of handling small cues that original SDM was not able to. One of the major difficulties encountered in using original SDM as an associative memory, is its inability to recover associations based upon relatively small cues; whereas we humans do. For a typical SDM to converge, a sufficiently large portion of a previously written word must be presented to the memory as an address.

The SDMSCue enhanced version of SDM, allows for handling small input cues and overcoming these pitfalls. Such cues were beyond the scope of original SDM work.

The ability of SDMSCue to overcome the input cue length constraint in the original SDM model provides superior functionality for associative memory. It allows for association and matching based on small hints or input cues.

The recall results obtained for SDMSCue are –in general- superior to those of original SDM. The gain achieved is quite significant statistically as well as objectively.

7. FUTURE RESEARCH

More comparisons and tests of SDMSCue vs. original SDM in specific AI and cognition domains may be done. Applying SDMSCue as an

associative memory technique for some architectures as well as test beds; is under consideration.

BIBLIOGRAPHY

Anwar, Ashraf (1997) *TLCA: Traffic Light Control Agent.* Master Thesis, University of Memphis, TN, USA, Dec 1997.

Anwar, Ashraf (1999) *Sparse Distributed Memory with Evolutionary Mechanisms.* Proceedings of Genetic and Evolutionary Computation Conference Workshop (GECCO) 1999, p. 339-40.

Anwar, Ashraf, Dasgupta, Dipankar, and Franklin, Stan (1999) *Using Genetic Algorithms for Sparse Distributed Memory Initialization.* Proceedings of Congress on Evolutionary Computation (CEC99), Jul 1999.

Anwar, Ashraf, and Franklin, Stan (2003) *Sparse Distributed Memory for 鏠onscious Software Agents.* Cognitive Systems Research Journal, Dec 2003, v 4 n 4, p 339-54, UK: Elsevier.

Evans, Richard, and Surkan, Alvin. (1991) *Relating Number of Processing Elements in a Sparse Distributed Memory Model to Learning Rate and Generalization.* APL Quote Quad, Aug 1991 v 21 n 4, p 166.

Franklin, Stan. (1995) *Artificial Minds.* MIT Press.

Franklin, Stan. (1997) *Autonomous Agents as Embodied AI.* Cybernetics and Systems, special issue on Epistemological Issues in Embedded AI.

Franklin, Stan. (2001). *Automating Human Information Agents.* Practical Applications of Intelligent Agents, ed. Z. Chen, and L. C. Jain. Berlin: Springer-Verlag.

Franklin, Stan, and Graesser, Art. (1999). *A Software Agent Model of Consciousness.* Consciousness and Cognition v 8, p 285-305.

Franklin, Stan, Kelemen, Arpad, and McCauley, Lee. (1998) *IDA: A Cognitive Agent Architecture.* IEEE transactions on Systems, Man, and Cybernetics, 1998.

Glenberg, Arthur M. (1997) *What Memory is for?* Behavioral and Brain Sciences.

Hely, T. (1994) *The Sparse Distributed Memory: A Neurobiologically Plausible Memory Model?* Master's Thesis, Dept. of Artificial Intelligence, Edinburgh University.

Kanerva, Pentti, and Raugh, Michael. (1987) *Sparse Distributed Memory.* RIACS, Annual Report 1987, NASA Ames Research Center, Moffett Field, CA, USA.

Kanerva, Pentti. (1988a) *Sparse Distributed Memory.* MIT Press.

Kanerva, Pentti. (1988b) *The Organization of an Autonomous Learning System.* RIACS-TR-88, NASA Ames Research Center, Moffett Field, CA, USA.

Kanji, Gopal K. (1999) *100 Statistical Tests.* Sage Publications.

Karlsson, Roland. (1995) *Evaluation of a Fast Activation Mechanism for the Kanerva SDM.* RWCP Neuro SICS Laboratory.

Kosslyn, Stephen M., and Koenig, Olivier. (1992) *Wet Mind.* Macmillan Inc.

Kristoferson, Jan. (1995a) *Best Probability of Activation and Performance Comparisons for Several Designs of SDM.* RWCP Neuro SICS Laboratory.

Kristoferson, Jan. (1995b) *Some Comments on the Information Stored in SDM.* RWCP Neuro SICS Laboratory.

Loftus, Geoffrey, and Loftus, Elizabeth (1976) *Human Memory, The Processing of Information.* Lawrence Erlbaum Associates.

Rao, Rajesh P. N., and Fuentes, Olac. (1996) *Learning Navigational Behaviors using a Predictive Sparse Distributed Memory.* Proceedings of the Fourth International Conference on Simulation of Adaptive Behavior, p 382.

Rao, Rajesh P. N., and Fuentes, Olac. (1998) *Hierarchical Learning of Navigation Behaviors in an Autonomous Robot using a Predictive Sparse Distributed Memory.* Machine Learning, Apr 1998 v31 n 1/3, p 87.

Rogers, D. (1988a) *Kanerva's Sparse Distributed Memory: An Associative Memory Algorithm Well-Suited to the Connection Machine.* International Journal of High Speed Computing, p 349.

Rogers, D. (1988b) *Using data tagging to improve the performance of Kanerva's Sparse Distributed Memory.* RIACS-TR-88.1, NASA Ames Research Center, Moffett Field, CA, USA.

Ryan, S., and Andreae, J. (1995) *Improving the Performance of Kanerva's Associative Memory.* IEEE Transactions on Neural Networks 6-1, p 125.

Scott, E., Fuller, C., and O'Brien, W. (1993) *Sparse Distributed Associative Memory for the Identification of Aerospace Acoustic Sources.* AIAA Journal, Sep 1993 v 31 n 9, p 1583.

Vogt, W. Paul (1998) *Dictionary of Statistics & Methodology, 2nd edition.* Sage Publications.

Willshaw, David. (1990) Coded Memories. A Commentary on 'Sparse Distributed Memory', by Pentti Kanerva. Cognitive Neuropsychology v 7 n 3, p 245.

TOWARDS A REALISTIC PERFORMANCE MODEL FOR NETWORKS OF HETEROGENEOUS COMPUTERS

Alexey Lastovetsky
and John Twamley

Abstract This paper presents experimental work undertaken towards the development of a realistic performance model for non-dedicated networks of heterogeneous computers. Unlike traditional models, it is aimed at parallel computing on common networks of computers and distributed computing on global networks. It takes into account the effect of paging and differences in the level of integration into the network of computers, with the inevitable fluctuation in the workloads of computers in such networks. Some preliminary experimental work undertaken in support of the development of the model is briefly discussed. Based on the results of these experiments some key parameters of such a model are proposed.

Keywords: performance models, heterogeneous networks, global networks, parallel computing, distributed computing

1. Introduction

Networks of computers (NOCs) are the architecture increasingly used for high performance computing. Over the last 10 years many applications have been written to efficiently solve problems on local and global NOCs. Computers on this type of network are heterogeneous in that they have different architectures. Programmers seeking to write parallel applications for heterogeneous networks, where the goal is to optimize the execution time of that application, must partition and distribute computations and data unevenly to the computers on which the application will run, in proportion to their relative speeds. The absolute speed of a computer may be defined as the number of computational units of a benchmark or test application, executed in a given time.

Relative computer speeds. Some performance model must be employed to predict and represent the relative speeds of the computers. Performance models employed have normally described relative computer speeds using constant positive numbers. Each computers speed is represented using the ratio of its speed against the other computers. Various methods have been employed to obtain these ratios. Early performance models used the MIPS (millions of instructions per second) or MFLOPS (millions of floating point operations per second) rates as the basis for the prediction of the comparative speed of computers. Another approach is to run the same benchmark application on each computer and compare their relative speeds. Relative computer speeds were used in [1] as an aid to partitioning 2D grids representing science and engineering problems and in [2] and [3] in partitioning Mmtrix multiplication problems. The use of real applications to obtain comparative computer speeds was used in the design of the mpC programming language [4].

Computer performance may be determined before execution, or dynamically, at run time. The best known system for use by dynamic schedulers to predict the performance of networked computers is the Network Weather Service [5]. The Network Weather Service monitors the fraction of the CPU utilization available to a newly started process on networked computers and represents this using decimal numbers. Available physical memory is also monitored, as well as network conditions. However, no allowance is made for differences in applications, or for the fluctuations in workload observed in networked computers as discussed below. Some examples of dynamic global based scheduling solutions can be found in [5],[6], [7] and [8].

In existing performance models these relative speeds are usually taken to be constant across a range of dataset sizes. This assumption is problematic; it has been shown that relative speeds can change with increased dataset sizes. If the computers on the network have significant differences in the size of each level of their memory hierarchies then their relative speeds will change as the size of the dataset used increases. The result of an experiment conducted by [9] using a serial application which multiplies two matrices is reproduced below. Table I shows the specifications of the computers used. Figure 1 illustrates the non-constant relative speeds exhibited as the application is run with increasing size datasets.

As the relative performance of the computers is not constant for all datasets, we cannot run a trial run of a task on each computer in the network with a small dataset, in order to accurately predict the relative performance of computers executing the task when run with a larger dataset. A model to accurately predict relative computer speeds for a particular task must reflect the fact that relative speeds will change as dataset sizes increase.

Non-dedicated networks and fluctuations in speed. Computers on NOCs used for high performance computing will experience fluctuations in their workload. This is a disadvantage of integration into the network. The higher the level of integration the greater the level of fluctuations observed. This changing transient load will cause a fluctuation in the speed of computers on the

Table 1. Specifications of 4 computers with different memory hierarchies

Machine. Name	*Architecture*	*CpuMHz*	*MainMemory*	*Cache(KB)*
Comp1	Linux 2.4.18-3 i686	499	513960	512
Comp2	SunOS 5.8 Ultra 5.10	440	409600	2048
Comp3	Linux 2.4.18-3 i686	996	254576	256
Comp4	Linux 2.4.18-3i686	499	126176	312

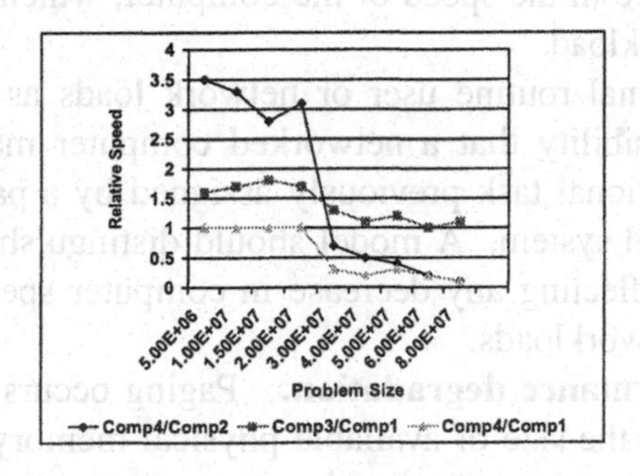

Figure 1. Non-constant relative speeds for computers with different memory hierarchies

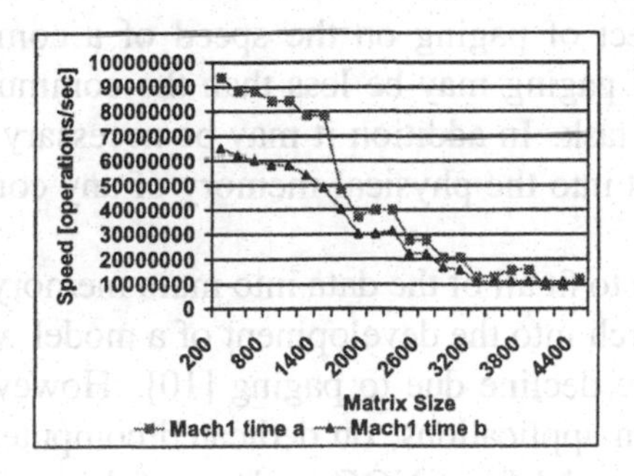

Figure 2. Differences in speed on a computer with a high level of network integration

network, in that the speed of a computer will vary when measured at different times while executing the same task.

Performance models have up to this point concentrated on partitioning data and computations on dedicated computer systems; they have not considered the issue of load fluctuation as a factor in the prediction of the computational speed of a networked machine. This fluctuation in workload on a computer is unpredictable; therefore it is impossible to make an accurate prediction of the speed of the computer for a particular application. A model for a non-dedicated heterogeneous network must take account of this. A realistic approach accepts that the speed of a computer for an application will be between an upper and lower level. In general the less integrated a computer is in a network the more accurate a speed prediction is likely to be. Figure 2 illustrates fluctuations in the speed of a server computer with a high level of network integration, running a serial matrix multiplication application. Two performance curves are shown, **time a** and **time b**, generated at different times. As can be seen there is a considerable difference in the speed of the computer, which can be explained by fluctuations in workload.

In addition to normal routine user or network loads as described above, there is also the possibility that a networked computer may be performing some heavy computational task previously assigned by a parallel application or scheduled by a grid system. A model should distinguish between the two types of workload, reflecting any decrease in computer speed resulting from heavy computational workloads.

Paging and performance degradation. Paging occurs when the dataset size of a task exceeds the size of available physical memory. A high level of paging will lead to a significant decline in a computers speed. In the past, the partitioning of parallel applications for dedicated systems was normally done to avoid paging. As a result, performance models have mostly assumed a linear relationship between problem size and execution time. However it is necessary to understand the effect of paging on the speed of a computer executing a task. The overhead of paging may be less than the communication overhead involved in dividing a task. In addition it may be necessary to remotely solve a task which cannot fit into the physical memory of any computers registered to solve the problem.

While it is desirable to fit all of the data into main memory, there was a need to conduct some research into the development of a model which does attempt to address performance decline due to paging [10]. However, this model assumes carefully written applications, on dedicated computer systems only. A realistic performance model for a NOC, to be useful in partitioning applications when subtasks may exceed the physical memory of available computers, must seek to model performance decline due to paging.

A new performance model for non-dedicated heterogeneous networks. Previous performance models provided good results for tasks which did not exceed physical memory on dedicated computer systems. However they are not likely to provide optimum results for scheduling subtasks which may exceed the memory of available computers on NOCs, where computers have different sizes at each level of their memory hierarchy. A realistic performance model for NOCs must take into account that relative computer speeds will not be constant, it may not be possible to accurately predict the speed of a computer for a particular application due to fluctuations in its workload and the application may not be suitable for partitioning to prevent paging. In addition, constraints on available computers or non-optimization in the design of the application may also lead to a situation where the avoidance of heavy paging is impossible.

The eventual goal of this research is to develop a performance model, capable of predicting the speed of a computer on a NOC, for a particular task. The model should take into account non-constant computer speeds, fluctuations in workload and paging, with an acceptable level of efficiency and accuracy. It should provide for improved execution times over existing models for parallel applications on NOCs, when used as a basis for determining computer speeds when allocating subtasks to available computers.

To the best of our knowledge there has been no experimental study conducted providing detailed data on how the speed of computers for a range of tasks on a non-dedicated heterogeneous network changes for increased datasets. In addition no measurements have been taken as to the effect of fluctuations in workload on the speed of computers in this type of network. Based on some basic preliminary experiments we have some ideas of the characteristics of the performance model. To prove that the model is applicable to a wide range of applications it is necessary to carry out extensive experiments using a wide range of applications, run on different computers.

Part 2 of this paper describes the initial requirements of a model. Part 3 describes the experimental study. It details the types of applications chosen as providing a good basis for inclusion in a representative set of tasks and the computers chosen to demonstrate different levels of network integration and specifications. Part 4 presents some preliminary results of experiments on our set of tasks, examining the shape of the curves generated and the effect that a varying load on non-dedicated computers has on the level of accuracy of any performance prediction. Performance bands are illustrated to represent this fluctuation. In part 5 we look at some conclusions detailing the potential of our model to fulfill some of the requirements outlined in part 2 and mention further work to be carried out in the area.

2. The model and motivations

Based on preliminary experiments our assumptions of the main features of the model can be summarized as follows.

- In order to take account of paging the model proposes to use a function to represent the performance of each individual computer, absolute speed against the size of the problem.

- The efficiency of the model is crucial. Computational and communication overheads for the building and maintenance of this model should not lead to unacceptable degradation in the overall performance. This can be achieved by the specification of particular cases, which may be applied where there is no loss of accuracy.
- The speeds of integrated computers on NOCs should be characterized by bands of curves and not by single curves. This reflects the reality that we cannot determine speed with absolute certainty, due to the unpredictable nature of workload fluctuations, but may be able to predict likely upper and lower speeds, giving a more realistic level of accuracy.
- The efficiency of the model might be increased by classifying computers by their level of network integration. Different computers may be treated differently to provide accuracy with good efficiency.
- The model must be effective for computers which are already executing a significant heavy computational workload. Computers in a network may be required to carry out more than one parallel or distributed computing task simultaneously.
- The model should be generic, applicable to both distributed computing scheduling and parallel computing partitioning tasks.

3. Experimental Study

In view of the initial requirements of the model as stated above, the following experimental strategy was followed.

- Experiments were conducted with tasks ranging from those with efficient memory reference patterns, to those with very inefficient reference patterns, in order to include tasks with wide differences in the rate of decline of the performance curves generated with increasing datasets.
- On examination of the generated performance curves, a number of tasks were selected as a representative set. This set provided a basis for the possible classification of tasks as approximations of one of this set, based on the rate of decline of its performance curve.
- By repeatedly running the set of tasks on computers at different times it was possible to determine the effect of a computers level of network integration on the speed of the computer. The fluctuations in speed were modelled as a performance band.
- The performance bands for our applications were examined to determine how a consideration of the level of network integration of the computer might lead to their more efficient generation.

Table 2. Computers and Specifications

Machine. Name	*Architecture*	Cpu MHz	*Memory*	*Cache(KB)*
Mach1	Linux 2.4.20-20.9bigmem i686	2783	7933500	512
Mach2	SunOS 5.8 Ultra 5.10	440	4096000	2048
Mach3	Windows XP	3000	1030388	512
Mach4	Linux 2.4.7-10i686	730	254524	256

- Some experiments were conducted with applications run on computers already executing a significant computational load in order to examine the effect on the characteristics of the performance band.

The applications chosen included those which displayed both the most gradual and the most severe decline in the performance curve generated with increasing datasets. The applications chosen were.

- ArrayOpsF. Arithmetic operations on 3 arrays of data, accessing memory in a contiguous manner, with an efficient memory referencing pattern and use of cache.
- TreeTraverse. Recursively traverses up a binary tree in a post-order manner. The contents of an array of integers are placed into a binary tree, resulting in non-contiguous storage of an equivalent number of nodes, each containing storage for an integer and a pointer to a child node. Memory is accessed in a random manner, with no benefits derived from caching.
- MatrixMultATLAS. An efficient implementation of matrix multiplication utilizing the cblas_dgemm BLAS routine, optimized using ATLAS.
- MatrixMult. A naive serial implementation of the multiplication of 2 N*N dense matrices with the result placed in a third matrix. Memory will be accessed in a non efficient manner with an increasing number of page faults over larger problem sizes resulting in heavy IO.

Table II provides the specifications of the computers used in our experiments. They were chosen to provide varying specifications and levels of network integration. They are representative of the range of computers typically found on a NOC.

Mach1 has a very high level of network integration. It is a computer science departmental server running NFS and NIS, as well as web and database servers, with a very high level of use and multiple users. Mach2 has a medium level of network integration, running database server and a web server. Mach3 runs p2p file sharing software and is connected up to the college LAN using Novelle, which provides a low level of integration. Mach4 is a networked machine, but provides no network services. It was anticipated that a high level of network integration would lead to a wide fluctuation in the speed of a computer over time.

4. Preliminary Experimental Results

4.1 Characteristics of performance curves

For each of the 4 applications, memory was obtained using malloc and gcc optimization -03 was used to compile the applications. It was considered that efficient memory referencing patterns would lead to a less steep decline in the speed of the application as problem sizes increased.

Figure 3 shows ArrayOpsF run on 4 computers. The performance curve generated by Mach1 is close to constant up to the maximum problem size capable of being solved. There is no sudden performance decline due to paging. Mach2, although its speed is significantly slower, exhibits the same constant characteristic. Mach3 displays the best performance until there is a sharp decline in speed, when paging causes the speed to dramatically decrease. A similar pattern is observed on Mach4, although the initial speed is slower than on Mach3, and the decline due to paging happens at a much smaller problem size. With the efficient memory access patterns and use of cache, the shape of the performance curve on all computers, except where paging causes a sudden decline in performance is close to constant. Figure 4 illustrates the execution speeds for 4 computers running TreeTraverse. Mach3 and Mach4 page heavily when their physical memory is exhausted. Mach1 levels off and becomes close to constant at problem sizes above 30 million nodes. Mach2 displays a slowly declining linear function. Figure 5 shows the generated curve for MatrixMultATLAS on our 3 UNIX clones. Mach1 and Mach2 are again close to constant for any significant problem size, Mach4 displays the same characteristic, but there is a decline in performance as paging occurs. Figure 6, because of the naive implementation of the application, was expected to show the most dramatic decrease in execution speed as problem sizes increased. Mach1 and Mach2 displayed a steady decline in speed with increasing problem sizes. Mach3 provided the fastest rate of execution over the whole of its problem range. Mach 4 also displayed a steady, reasonably linear decline. The experiments revealed that Mach1 and Mach2 are configured to avoid paging. This is typical of computers used as a main server. For applications designed to efficiently use cache memory, such computers may be characterized by a constant stepwise function, up to the point where the process crashes, probably because it tries to invoke a paging procedure, not allowed due to its configuration. This

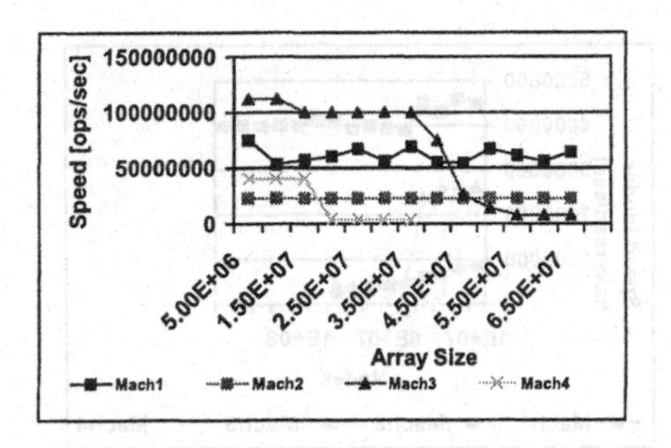

Figure 3. ArrayOpsF on 4 computers

is illustrated by examining the performance of Mach1 and Mach2, running MatrixMultAtlas, figure 5, with its efficient use of cache. Apart from small problem sizes, the speeds of both computers are constant. In contrast, both computers running MatrixMult, figure 6, show a decreasing speed with increasing problem sizes, levelling out only with large dataset sizes. This slowdown cannot be explained by paging, but is due to an increasing number of cache misses as the problem size increases. Non-paging computers, running tasks which reference memory in a random manner, not benefiting from caching, may also be characterized by a constant stepwise function. This is illustrated by figure 4, where we observe a close to constant speed for Mach1 and Mach 2, for all significant problem sizes, running TreeTraverse.

The efficiency of the overall performance model may be improved by differentiating between paging and non-paging computers, running applications designed to efficiently use cache, or applications which reference memory in a random manner. Although these 2 applications display contrasting levels of efficiency of cache use, they are similar in that the contribution of caching to performance is consistent for all problem sizes. Where computers are configured to prevent paging, the use of a constant stepwise function offers considerable efficiency gains.

Figure 7 illustrates the effect of cache use efficiency and paging on our performance curve, for each of our 4 applications, on a machine where heavy paging is permitted (Mach3 and Mach4). Performance is characterized using piecewise linear functions. The influence of caching and paging on the shape of the performance curve is illustrated for each application. The shape of the curve depends only on tasks and looks similar for both paging computers. For small dataset sizes, capable of being accommodated in physical memory, the slope of the curve is determined by the contribution of caching to performance. Where it is constant, as occurs with optimized tasks and tasks which do not

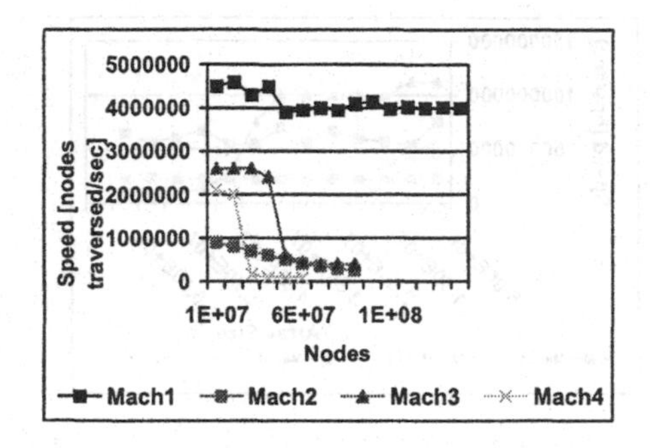

Figure 4. TreeTraverse on 4 computers

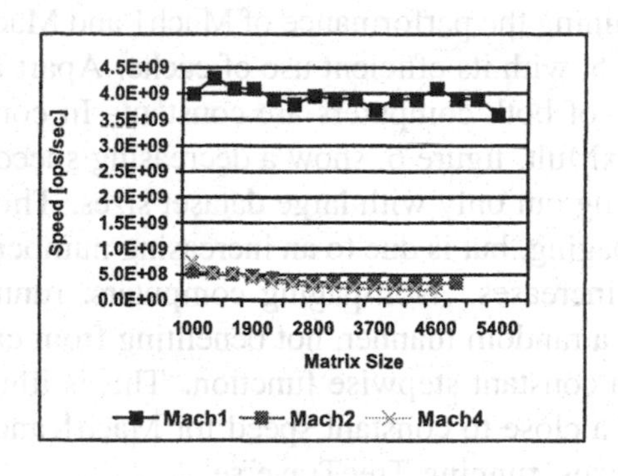

Figure 5. MatrixMultATLAS on 3 computers

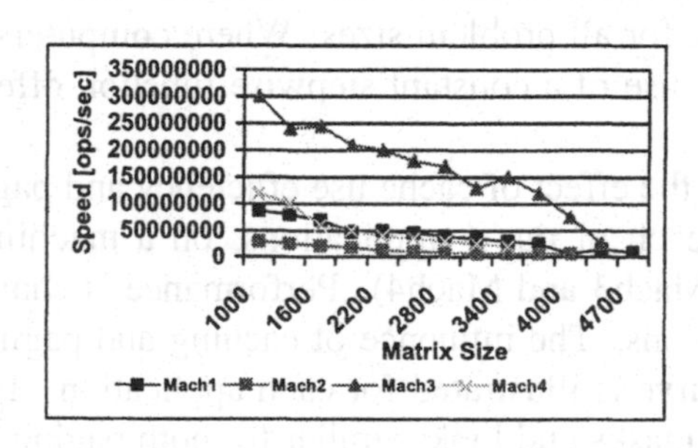

Figure 6. MatrixMult on 4 computers

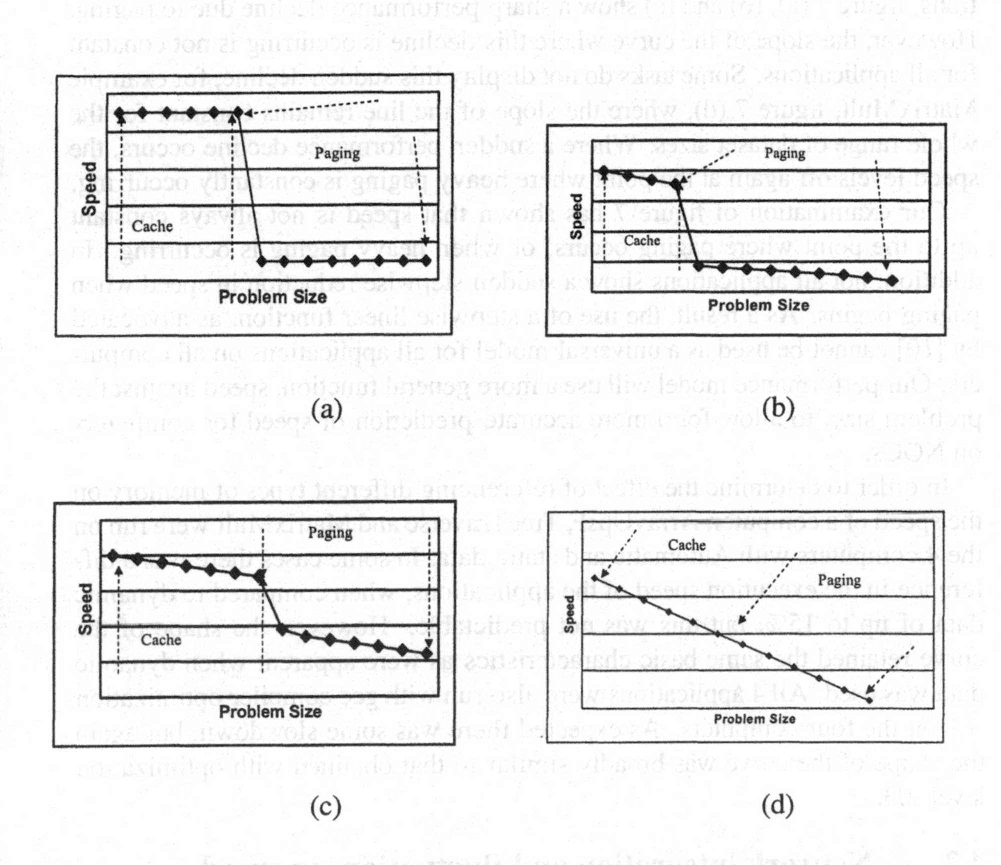

Figure 7. The effect of caching and paging in reducing the execution speed of each of our 4 applications, with performance characterized using piecewise linear functions .

benefit to any extent from caching, the slope of the curve will not deviate too significantly from the horizontal. This is apparent in figure 7 (a) and figure 7 (c), where the applications benefit from caching, and in figure 7 (b), where random memory reference patterns do not allow any benefit from caching. Figure 7 (d) illustrates an application where there is an increase in the proportion of cache misses, with increasing problem sizes. We notice a decline in speed for problem sizes smaller than those where paging occurs. Three of our applications, figure 7 (a), (b) and (c) show a sharp performance decline due to paging. However, the slope of the curve where this decline is occurring is not constant for all applications. Some tasks do not display this sudden decline, for example MatrixMult, figure 7 (d), where the slope of the line remains constant for the whole range of dataset sizes. Where a sudden performance decline occurs, the speed levels off again at the point where heavy paging is constantly occurring.

Our examination of figure 7 has shown that speed is not always constant up to the point where paging occurs, or when heavy paging is occurring. In addition, not all applications show a sudden stepwise reduction in speed when paging begins. As a result, the use of a stepwise linear function, as advocated by [10] cannot be used as a universal model for all applications on all computers. Our performance model will use a more general function, speed against the problem size, to allow for a more accurate prediction of speed for computers on NOCs.

In order to determine the effect of referencing different types of memory on the speed of a computer, ArrayOpsF, TreeTraverse and MatrixMult were run on the 4 computers with Automatic and static data. In some cases there was a difference in the execution speed of the applications, when compared to dynamic data of up to 15%, but this was not predictable. However, the shape of the curve retained the same basic characteristics as were apparent when dynamic data was used. All 4 applications were also run with gcc compiler optimization -01 on the four computers. As expected there was some slowdown, but again the shape of the curve was broadly similar to that obtained with optimization level -03.

4.2 Network integration and fluctuations in speed

Having considered the performance of the set of 4 applications on a range of computers typically found on NOCs, it is now necessary to examine the effect of a computers level of network integration on the accuracy of speed prediction. As stated above, a performance band instead of a curve was thought to offer an appropriate mechanism to model the realities of workload fluctuation. The performance band may be defined as the level of fluctuation which may occur in the performance of a computer executing a particular problem, due to changes in load over time, expressed as a percentage of the maximum speed of execution for that problem.

The routine load on the computers was monitored and the 4 applications were run over a range of system loads. Figure 8 illustrates the performance band of our 4 computers running MatrixMult and TreeTraverse. Figure 8 (a) illustrates that the performance band for MatrixMult on Mach1 is around 40% for smaller problem sizes, narrowing to around 6% for larger problems. The decrease in absolute speed with increased problem sizes is due to the effects of caching, the computer is configured to avoid paging. For problem sizes with much longer execution times load fluctuations will be averaged out leading to a narrowing of the performance band. Figure 8 (b) shows the performance band for TreeTraverse on the same machine. The execution time is shorter than MatrixMult, so we don't see a narrowing of the band. In this case the performance band is in the order of 35% for it's whole range of problem sizes. Figure 8 (c) illustrates the performance band for Mach2 running MatrixMult. It can be seen that for smaller problem sizes the performance band is in the order of 15%. As might be expected the load fluctuation on this machine was less than that present on Mach1. Again the decline in absolute speed can be attributed to caching. The performance band displayed by Mach2 for TreeTraverse as shown in figure 8 (d) is quite small, being reasonably constant at 8% for anything other than the smallest problem sizes. The performance band displayed by Mach3 for MatrixMult and TreeTraverse, shown in figure 8 (e) and (f), with it's low level of network integration, was not greater than around 5 to 7% even when there was heavy file sharing activity (its effect seemed minimal). Mach4 displayed very little fluctuation for our 2 applications, with a performance band of around 3 to 5%, as might be expected from a virtually stand alone computer.

The size of the performance bands for Mach1 and Mach2, the computers with a high level of network integration, against time, are shown in figure 9. The influence of workload fluctuations on speed becomes less significant as the execution time increases. There is a close to linear decrease in the size of the performance band as the execution time increases.

We have noted that the fluctuation in speed observed for Mach1, the computer with the highest level of integration, is in the order of 40% for small problem sizes, declining to approximately 6% for the maximum problem size solvable on the computer. This level of fluctuation justifies the use of bands to model the speed of computers, instead of curves. The accuracy obtained with bands can still be within acceptable limits to be useful when predicting performance. The level of accuracy of the approximation is increased as the execution time increases with larger problem sizes. For computers with a low level of network integration such as Mach4 performance bands allow for a high level of accuracy in the prediction of performance.

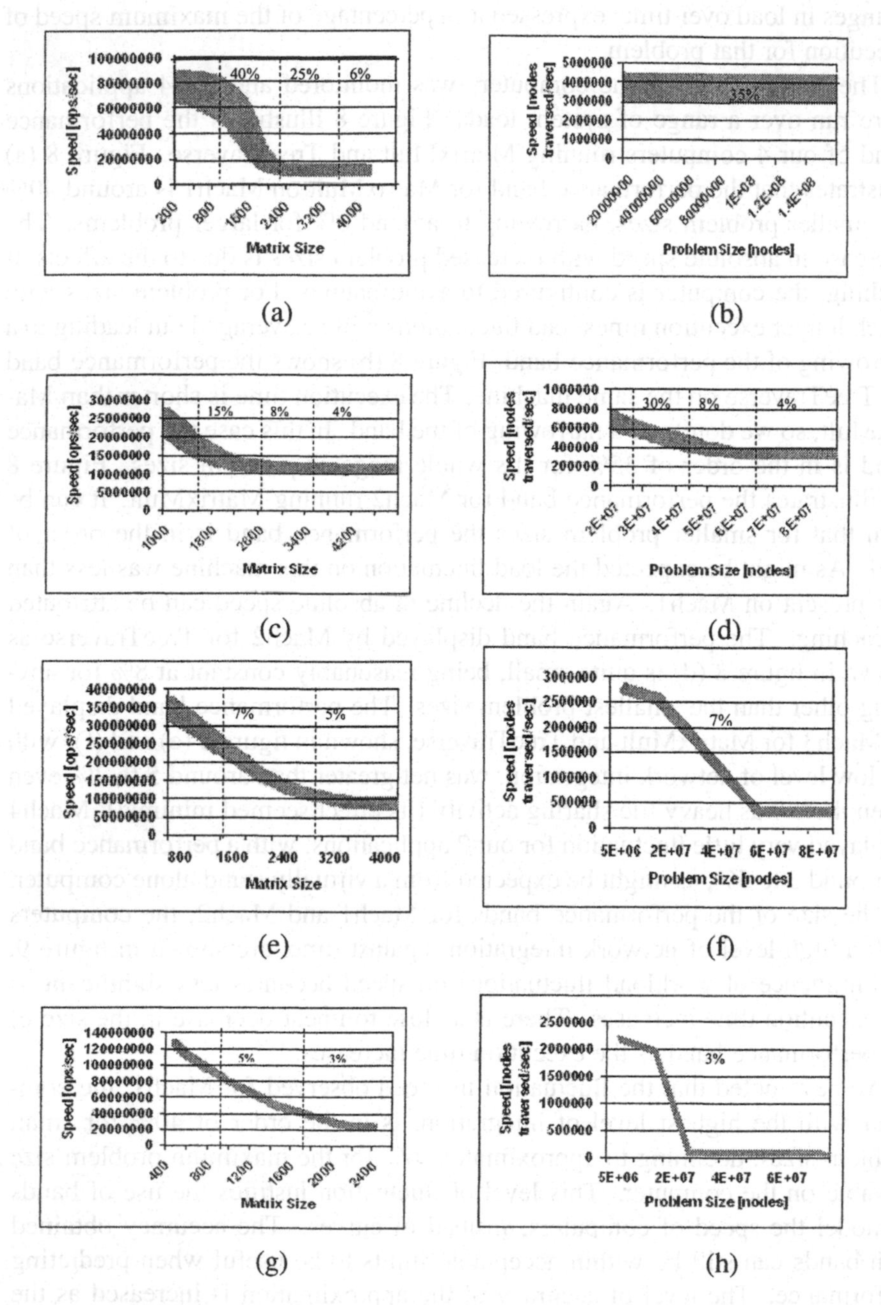

Figure 8. (a) Performance band for MatrixMult on Mach1, (b) Performance band for TreeTraverse on Mach1, (c) Performance band for MatrixMult on Mach2 , (d) Performance band for TreeTraverse on Mach2, (e) Performance band for MatrixMult on Mach3, (f) Performance band for TreeTraverse on Mach3, (g) Performance band for MatrixMult on Mach4, (h) Performance band for TreeTraverse on Mach4.

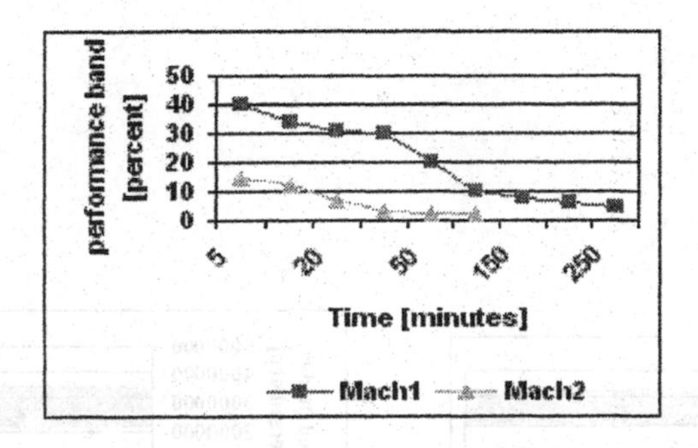

Figure 9. Performance band as a function of time on Mach1 and Mach2.

We have established that a different level for the accuracy of prediction is achievable for computers with high and low levels of integration. The efficiency of the model can be increased by taking this into account and treating computers differently. The number of measurements of speed against problem size required to achieve a realistic level of accuracy is inversely proportionate to the level of integration. A computer with a high level of integration will require very few measurements. A less integrated computer will require more measurements to reflect the higher level of accuracy achievable. The increased computational overhead required by computers with a low level of integration is offset by the likelihood of many of these computers being present on a typical network, with the prospect of utilizing a significant computational resource.

It is possible that computers displaying good performance, may be assigned, in addition to their routine workload, more than 1 heavy computational task. The performance band of Mach1, a computer with 4 processers, running Tree-Traverse, while already processing 4 substantial computational loads was constructed, simulating a situation where additional heavy loads are being executed on a computer. The effect of the heavy computational load on the performance band of the computer was determined. Figure 10 (a) shows the performance band while the workload was restricted to normal fluctuating routine computations. Figure 10 (b) shows the performance band generated while the additional 4 heavy computational loads were being run on the computer. Both the upper and lower level of speed is reduced but the overall width of the band remains relatively constant. It is thought likely that this lowering of the performance band occurs on all computers, executing all tasks, where the number of prior computational loads is greater than or equal to the number of processors present on the computer. More experimental work is required to investigate the effects of heavy existing workloads on performance predictions, particularly for single processor computers.

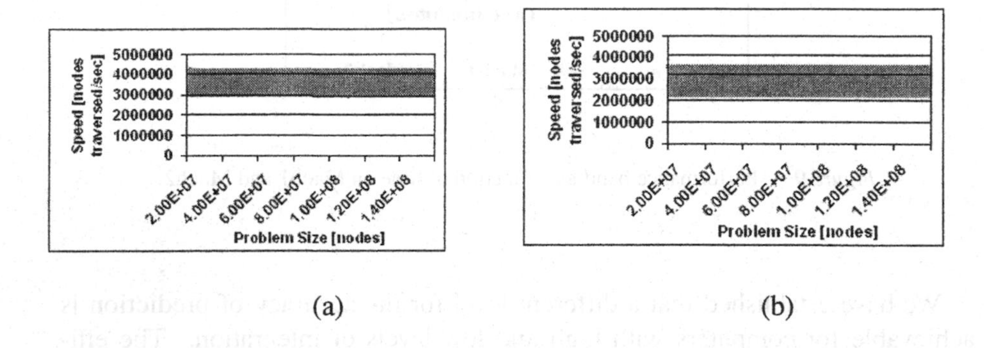

Figure 10. (a) Performance band for TreeTraverse on Mach1 with normal fluctuating workload. (b) Performance band for TreeTraverse on Mach1 with 4 prior major computational loads.

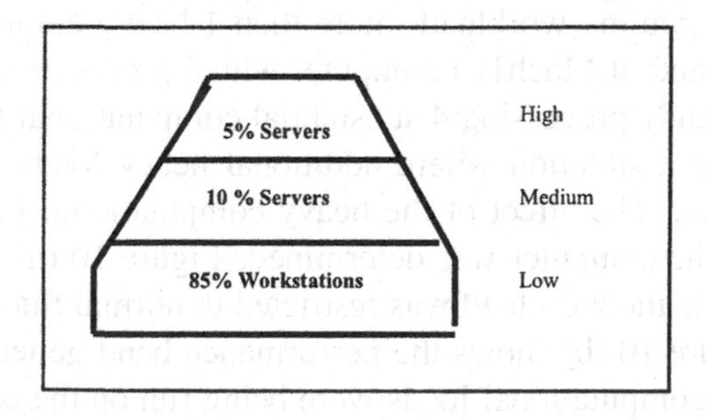

Figure 11. Typical avaliability of machines on a network classified by network integration and function

5. Conclusions and further research

Based on the experimental results we can summarize the primary features necessary for inclusion in the model

- To address paging in parallel and high performance grid computing we use a function, the size of the problem against the absolute speed of the computer, rather than a constant or stepwise function. We have observed that the slope of the performance curve generated by different tasks may vary considerably. Also, for some tasks there is no sudden decline in speed when paging occurs.
- The model should be composed of sub-models, thereby increasing the efficiency of the overall model. The sub-model best suited to model the speed of a computer executing a particular application should be dependent on the computers level of network integration. Our approach is to seek to maximize efficiency where possible, without decreasing the accuracy of prediction. For example, a computer with a high level of integration will require less measurements to construct its performance band to an acceptable level of accuracy.
- The use of bands of curves, instead of a single curve, is the best method to model performance on NOCs. The width of the performance bands observed on computers with a high level of network integration is still sufficiently narrow to allow for an acceptable estimation of performance.
- Computers should be classified according to their level of network integration. Computers with a high level of network integration show large performance bands for small problem sizes, narrowing as the execution time increases. Initial observations indicate that the performance band may be approximated to a linear decrease over time, but more experimental work is needed to investigate this observation.
- The model should classify all computers on a network. Figure 11 shows the breakdown of computers on a typical NOC. The computational resource available when we combine our loosely integrated computers is significant and will usually exceed that of our highly integrated computers, therefore we must understand their performance characteristics if we are to efficiently exploit available resources.
- In our model we differentiate between computers that allow paging and computers that do not. Non-paging computers are typically servers with a heavy workload, offering a good relative performance. For carefully designed applications, efficiently utilizing cache, and for applications where memory is referenced randomly, the performance band of this

type of computer is linear, up to the maximum size of the problem solvable on the computer.

Some preliminary experimental work suggests its possible to reflect the fact that a computer is already engaged in a heavy computational task when seeking to predict its speed for a particular task. Initial work has suggested that while the upper and lower levels of speed are reduced, the width of the performance band remains close to constant. However, it may be possible to distinguish 2 types of prior computational load, each of which will have a different effect on the characteristics of the performance band. A heavy computational load with a small dataset may narrow the performance band, but not influence the dataset size at which heavy paging begins. In contrast, a task which utilizes a significant percentage of physical memory will cause paging to occur for even a small dataset. It is likely for some tasks, that paging may occur for even the smallest datasets. It may be possible, based on a measurement of CPU utilization and the degree of paging observed to predict the likely effect on the performance band. This would increase the efficiency of the model where heavy computational work is being executed on a computer. However, further work is required to investigate the feasibility of this approach.

More experimental study will be carried out to confirm the preliminary experimental results presented in this paper. The goal is to provide a practical performance model for NOCs enabling a realistic prediction of the speed of a computer for a particular task.

References

[1] Crandall, P. and M. Quinn. Problem Decomposition for Non-Uniformity and Processor Heterogenity. *Journal of the Brazillian Computer Society*, 2(1):13–23, 1995.

[2] Kalinov, A. and A. Lastovetsky. Heterogeneous Distribution of Computations While Solving Linear Algebra Problems on Networks of Heterogeneous Computers. *Proceedings, High Performance Computing and Networking Europe 1999*, pp. 191-200, 1999.

[3] Beaumount, O., Boudet, V., Rastello, F. and Y. Robert. Matrix Multiplication on Heterogeneous Platforms. *IEEE Transactions on Parallel and Distributed Systems*, 12(10):1033–1051, 2001.

[4] Arapov, D., Kalinov, A., Lastovetsky, A. and I. Ledovskih. A Language Approach to High Performance Computing on Heterogeneous Networks. *Parallel and Distributed Computing Practices*, 2(3):323-332, 1999.

[5] Wolski, R. Dynamic Forecasting Network Performance Using the Network Weather Service. *Cluster Computing*, (1) :119–132, January 1998.

[6] Zaki, M.J., Lei, W. and Parthasarathy, S. Customized Dynamic Load Balancing for a Network of Workstations. *Journal of Parallel and Distributed Computing*, 43, 156-162, 1997.

[7] Berman, F. *High Performance schedulers. The Grid: Blueprint for a New Computing Infrastructure*. Morgan-Kaufmann, 1988.

[8] Vadhiyar, S. , Dongarra, J. and A. Yarkhan. GrADSolve - RPC for High Performance Computing on the Grid. *Euro-Par 2003, 9th International Euro-Par Conference, Proceedings*, Springer, LCNS 2790, pp 394-403, August 26-29, 2003

[9] Lastovetsky, A. and R. Reddy. Data Partitioning with a Realistic Model of Networks of Heterogeneous Computers. *Proceedings of the 18th International Parallel and Distributed processing symposium* (IPDPS 2004), 26-30 April 2004, Santa Fe, New Mexico, USA. IEEE Computer Society Press.

[10] Drozowski. M. Out-of-Core Divisible Load Processing. *IEEE Transactions on Parallel and Disbributed Computing*, 14(10):1048-1056, 2001.

EXTENDING CLUSTERSIM WITH MP AND DSM MODULES

Christiane Pousa, Luiz E. Ramos, Luís F. Góes, Carlos A. Martins

Graduation Program in Electrical Engineering, Pontifical Catholic University of Minas Gerais, Belo Horizonte, Minas Gerais, Brazil

Abstract: In this paper, we present a new version of ClusterSim (Cluster Simulation Tool), in which we included two new modules: Message-Passing (MP) and Distributed Shared Memory (DSM). ClusterSim supports the visual modeling and the simulation of clusters and their workloads for performance analysis. A modeled cluster is composed of single or multi-processed nodes, parallel job schedulers, network topologies, message-passing communications, distributed shared memory and technologies. A modeled workload is represented by users that submit jobs composed of tasks described by probability distributions and their internal structure (CPU, I/O, DSM and MPI instructions). Our **main objectives** in this paper are: to present a new version of ClusterSim with the inclusion of Message-Passing and Distributed Shared Memory simulation modules; to present the new software architecture and simulation model; to verify the proposal and implementation of MPI collective communication functions using different communication patterns (Message-Passing Module); to verify the proposal and implementation of DSM operations, consistency models and coherence protocols for object sharing (Distributed Shared Memory Module); to analyze ClusterSim v.1.1 by means of two case studies. Our **main contributions** are the inclusion of the Message-Passing and Distributed Shared Memory simulation modules, a more detailed simulation model of ClusterSim and new features in the graphical environment.

Key words: Cluster Computing, Discrete-Event Simulation, Distributed Shared Memory, Message-Passing Communications, Performance Analysis

1. INTRODUCTION

Nowadays, clusters of workstations are widely used in academy, industry and commerce. Usually built with "commodity-off-the-shelf" hardware components and freeware or shareware available in the web, they are a low cost and high performance alternative to supercomputers[1,2]. The performance analysis of different DSM software, message passing libraries, parallel job scheduling algorithms, consistency models, interconnection networks and topologies, heterogeneous nodes and parallel jobs on real clusters is complex. It requires: a long time to develop and change software; a high financial cost to acquire new hardware; a controllable and stable environment; less intrusive performance analysis tools etc. On the other hand, analytical modeling for the performance analysis of clusters requires too many simplifications and assumptions[3].

Thus, simulation is a less expensive (financial cost) performance analysis technique than measurement on real systems. Besides, simulation is also more accurate than analytical modeling for evaluating the performance of a cluster and constructive blocks. It allows a more detailed, flexible and controlled environment. It makes feasible for researchers to compare a wide variety of clusters configurations and workloads[3,4]. Besides, there are many available types of tools that aid the development of simulations: simulation libraries (SimJava[5,6], JSDESLib[1] etc), languages (SIMSCRIPT[4], SLAM[4] etc) and application specific simulation tools (Gridsim[3], Simgrid[7], SRGSim[8], MPI-SIM[9], PP-MESS-SIM[10], DSMSim[11] etc).

In this paper, we present the implementation of the new version of Cluster Simulation Tool (ClusterSim v.1.1) that is used to model clusters and their workloads through a graphical environment, and evaluate their performance using simulation. The tool is an evolution of RJSSim[12] and ClusterSim v.1.0[13]. This new version was extended with two new modules: the Message-Passing Module (MP module) and the Distributed Shared Memory Module (DSM module). It also presents a more detailed simulation model and some new features in the graphical environment. The main objectives of this paper are: to present and verify both new modules (MP and DSM) of ClusterSim v.1.1.; to analyze ClusterSim by means of two case studies. The **main goals** of this paper are: the proposal and implementation of two new simulation modules for ClusterSim: MP and DSM; the simulation of MPI collective communication functions using different communication patterns (using MP); the proposal and implementation of DSM operations, consistency models and coherence protocols for object sharing (using DSM); the introduction of new features in the graphical environment; and an automatic workload generation module.

2. RELATED WORKS

In the last years, simulation has been used as a powerful technique for performance analysis of computer systems[14,15]. Researchers usually build simplified specific purpose simulation tools and do not describe them in detail or do not make available the source code and/or binaries. Nevertheless some works regarding the simulation of parallel, network, cluster and grid computing make available their simulation tools and documentation[3,7,8,9,10,11,16,17,18]. A detailed analysis of those works is found in our previous papers[1,14] and other simulation tools can be found in[18,19].

In this section, we briefly describe the simulation tools that have the closest relation with the new two modules of ClusterSim: a cluster computing simulation tool (SRGSim[8]), two message passing simulators (MPI-SIM[9] and PP-MESS-SIM[10]) and two DSM simulation tools (DSMSim[11] and Mica[20]), which present interesting features.

SRGSim is a Java-based discrete-event simulation tool developed in the University of California. It simulates some classes of parallel job scheduling (dynamic and static scheduling), architectures (cluster of computers, multiprocessors etc) and jobs through probabilistic models, constants and DAGs (Direct Acyclic Graphs). The main features of SRGSim are: a DAG editor, the description of jobs by means of traces, probabilistic models or DAGs, the use of CPU-, I/O- and Communication-bounded jobs, the support to some network topologies, and its parallel and distributed implementation. SRGSim has a text-only interface and does not support multithreading. Moreover, it does not support heterogeneous nodes and mechanisms that simulate the MPI communication functions[8].

MPI-SIM[9] is a library in C for the execution-driven parallel (multithreaded) simulation of MPI programs. The simulated workload can use a subset of available MPI point-to-point primitives and collective communication functions. MPI-SIM predicts the performance of the MPI programs as a function of architectural characteristics (number of processors, network latencies etc). The simulator supports point-to-point networks only.

PP-MESS-SIM[10] is a C++ object-oriented event-driven simulator for message passing in multicomputers. Its main features are: the support to different router architectures, the possibility of multiple coexisting routing policies and switching schemes, the extensibility, the representation of traffic classes and workloads as distribution functions and the use of an input language (a text-only interface).

DSMSim[11] is a C++- distributed shared memory simulation tool developed in the University of Wayne State. In DSMSim a workload can only be represented by real programs of String DSM system. The simulation model is based on events and entities. Only one consistency model and one coherence protocol were implemented. DSMSim implements a page-based DSM and decentralized management for shared pages. Besides, the tool does not have a graphical environment for modeling the cluster architecture and uses simple statistical analysis mechanisms.

MICA is a memory and interconnect simulation environment for cache-based architectures. It simulates DSMs that were implemented in hardware and uses real traces as workloads. The simulation is driven by events rather than by system clocks, which in turn speeds up the simulation whereas preserving accuracy. MICA was built on top of CSIM[20], which supports the use of a variety of scheduling policies and complex systems. Since CSIM provides user-level threads, it can simulate a set of computing nodes, point-to-point and collective communication activities through these threads. MICA provides: cache, bus, memory, directory, node controller/network interface and interconnection modules. It implements two page allocation policies and some routing algorithms[21].

In spite of the SRGSim has a detailed probabilistic workload simulation model, its model does not provide a detailed structural description of a job (ex: loop structures, different communications functions etc). Moreover, the workload probabilistic description is less representative, but it requires less simulation time. Thus, the structural description of the workload makes ClusterSim more representative. Besides, its graphical environment makes it easier to use. As well as SRGSim, ClusterSim implements CPU, I/O and communication functions that support the definition of different communications patterns and algorithms models with varying granularity.

ClusterSim simulates MPI collective communication functions with basis on point-to-point primitives like PP-MESS-SIM and MPI-SIM. However, MPI-SIM requires real C/C++/UC programs in order to run a simulation, whereas ClusterSim uses a hybrid language independent workload model. Besides, unlike ClusterSim and PP-MESS-SIM, MPI-SIM only simulates point-to-point networks. On the other hand, PP-MESS-SIM uses a specific language for workload definition, instead of a graphical environment.

As well as DSMSim and Mica, ClusterSim can simulate distributed shared memory (DSM). Nevertheless, Mica only simulates DSM implemented in hardware, which uses pages as grains (invariable granularity). ClusterSim simulates a variety of consistency models, access protocols, coherence protocols and some synchronization primitives that were not found in DSMSim. Both DSMSim and Mica, just accept real programs as input, unlike ClusterSim in which real and abstract DSM programs can be simulated.

3. CLUSTER SIMULATION TOOL (CLUSTERSIM)

ClusterSim is a Java-based parallel discrete-event simulation tool for cluster computing. It supports the visual modeling and the simulation of clusters and their workloads for performance analysis. A modeled cluster is composed of single or multi-processed nodes, parallel job schedulers, network topologies, distributed shared memory, message-passing mechanisms and technologies. A workload is represented by users that submit jobs composed of tasks described by probability distributions and their internal structure.

The main features of ClusterSim v.1.0 (old version) are: a graphical environment to model clusters and their workloads; an available source code; extensible classes; a mechanism to implement new job scheduling algorithms, network topologies, etc; representation of a job by some probability distributions and internal structure (loop structures, CPU, I/O, MPI (communication) instructions); support to the modeling of clusters and heterogeneous or homogeneous nodes; an independent thread for each simulation entity (architectures and users); the support to some collective and point-to-point MPI (Message Passing Interface) functions; representation of a network by its topology, latency, bandwidth, protocol overhead, error rate and maximum segment size; the support to different parallel job scheduling algorithms (space sharing, gang scheduling, etc) and node scheduling algorithms (first-come-first-served (FCFS), etc); a statistical and performance module that calculates some metrics (mean nodes utilization, mean simulation time, mean jobs response time etc); the support to some probability distributions (Normal, Exponential, Erlang Hyper-Exponential, etc) to represent the parallelism degree of the jobs and the inter-arrival time between jobs submissions and the possibility of specifying the simulation time and seed.

ClusterSim v.1.0 didn't have distributed shared memory and the message-passing functions were implemented with unique communication patterns. Besides, the user must model the message-passing workloads manually. By including a couple of modules (MP and DSM) the new features of ClusterSim v.1.1 are:

- The support to different communication patterns (ring, binomial tree, etc), memory consistency models (sequential, weak, etc) and coherence protocols (write-invalidate eager, write-update lazy, etc);
- A graphical environment for modeling clusters and workloads that use distributed shared memory;
- A graphical environment for the automatic modeling of message-passing workloads.

3.1 The Architecture of ClusterSim

The architecture of ClusterSim is divided into three layers: graphical environment, entities and core. The first layer allows the modeling and the simulation of clusters and their workloads by means of a graphical environment. Moreover, it provides statistical and performance data about each executed simulation. The second layer is composed of three entities: user, cluster and node. Those entities communicate by means of event exchange supported by the simulation core. The third layer is the simulation core itself, a discrete-event simulation library named JSDESLib[1].

3.1.1 Graphical Environment

The graphical environment was implemented using Java Swing and NetBeans 3.4.1 IDE. ClusterSim v.1.0 is composed of a configuration and simulation execution interface, three workload editors (user, job and task editors) and three architecture editors (cluster, node and processor editors). In the new version of ClusterSim, two new workloads editors (auto job and object editors) were implemented. Besides, some updates were implemented on the existing workload and architecture editors. Using these tools, it is possible to model, execute, save and modify simulation environments and experiments (Fig. 1). As well as ClusterSim editors, the simulation model is divided between workload and architecture.

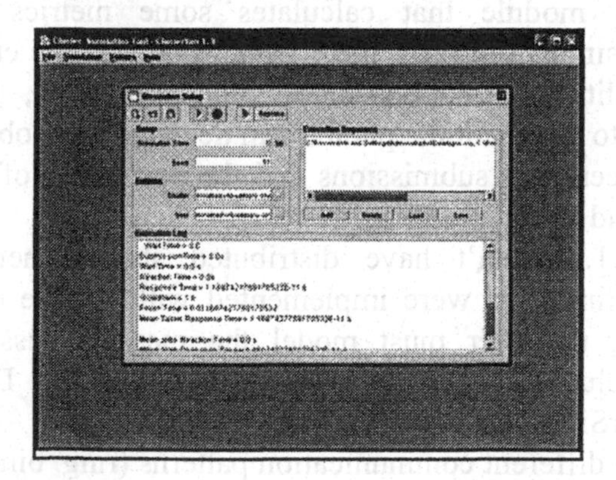

Figure 1. Main interface of the graphical environment

Based on the related works, we chose a hybrid workload model using probability distributions to represent some parameters (parallelism degree and inter-arrival time) and an internal structure description of the jobs. The execution time is a parameter that is commonly found in execution logs, but it is only valid for certain workloads and architectures. Besides, it is influenced by many factors such as: load, nodes processing power, network overhead etc. Thus, the execution time of a job must be calculated during a simulation, according to the simulated workload and architecture.

We model a job by describing its internal structure, instead of using its execution time. This approach has many advantages: i) real jobs can be represented directly (without structure modifications); ii) the execution time of a job is dynamically calculated and is only influenced by the simulated environment; iii) the job's behavior is more deterministic; iv) many parallel algorithm models and communication patterns can be implemented.

Unlike execution time, parallelism degree is usually known and indicated by the user that submits a job. As parallelism degree is not influenced by execution time factors, it can be represented by a probability distribution.

In order to avoid long execution traces, the jobs inter-arrival time is also represented by a probability distribution. Moreover, exponential and Erlang hyper-exponential distributions are widely used in the academic community to represent the jobs inter-arrival time[14,15].

In order to model a workload, we can use the following editors: User, Job or Auto Job, Objects and Task. In the User Editor, it is necessary to specify: the number of jobs to be submitted, the inter-arrival time distribution and the jobs types. The submission probability must be specified for each job type. The sum of those probabilities must be equal to 100%. The jobs types are selected through Monte Carlo's method, where a random number between 0 and 1 is raffled, indicating the job to be submitted. For each new instance of a submitted job type, new values are sampled for the parallelism degree of each task and inter-arrival time.

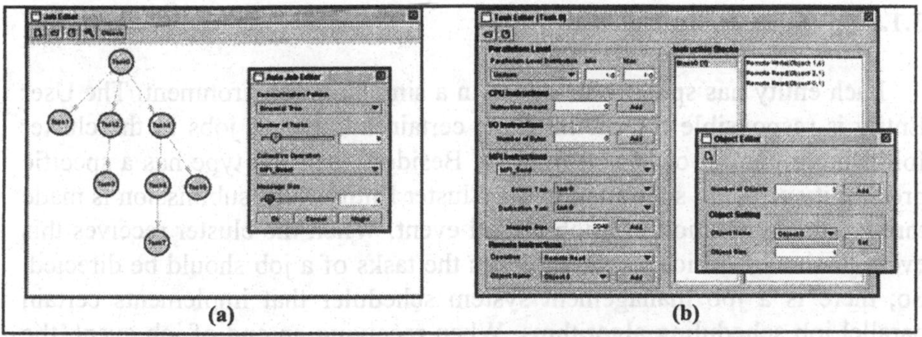

Figure 2. (a) Job and Auto Job editors; (b) Task and Objects editors

The Job Editor (Fig. 2a) allows the specification of a job through a graph. Each node represents a task and each edge represents the communication between two tasks. Starting from the Job Editor, it is possible to edit each task by activating the Task Editor (Fig. 2b) and create and edit objects for task instructions by activating the Objects Editor. In the Task Editor CPU,

E/S, DSM and MPI instructions are inserted into instruction blocks and the distribution of the parallelism degree is specified. Each instruction can automatically be parallelized according to the parallelism degree of the task. For instance, suppose a parallel job that follows the process farm model. It would be modeled with two nodes: the master task (parallelism degree equal to 1) and the slave tasks (parallelism degree equal to n). If the parallelization option was active in the CPU instructions of a slave, the total number of CPU instructions of each slave would be equal to the total number of CPU instructions divided by the parallelism degree n. Thus, it is possible to verify the speedup achieved by the parallelization of a job, without implementing a different job for each parallelism degree.

In order to aid the user on modeling workloads, we implemented an Auto Job Editor (Fig 2a), which automatically creates a job containing a user-defined number of tasks that communicate according to some available communication pattern. We intend to improve that editor in the following versions of ClusterSim.

3.1.2 Statistical and Performance Module

For each executed simulation, the statistical and performance module of ClusterSim creates a log with the calculation of several metrics. The main calculated metrics are: mean jobs and tasks response time; wait, submission, start and end time of each task; number of messages exchanged between nodes (DSM); transmission time (MP); overhead for memory copies; mean jobs slowdown; mean nodes utilization; mean jobs reaction time etc.

3.1.3 ClusterSim's Entities

Each entity has specific functions in a simulation environment. The User Entity is responsible for submitting a certain number of jobs to the cluster following a pattern of arrival interval. Besides, each job type has a specific probability of being submitted to the Cluster Entity. This submission is made through the generation of a job arrival event. When the cluster receives this event, it should decide to which nodes the tasks of a job should be directed. So, there is a job management system scheduler that implements certain parallel job scheduling algorithms. When receiving an end of job event, the cluster scheduler should remove the execution job. Fig. 3 shows the events exchange diagram of ClusterSim, detailing the interaction among the User, Cluster and Node entities. To simplify the diagram, some classes were omitted. Other important classes belonging to the cluster entity are: Single System Image, Network, MPI manager and Shared Memory Manager.

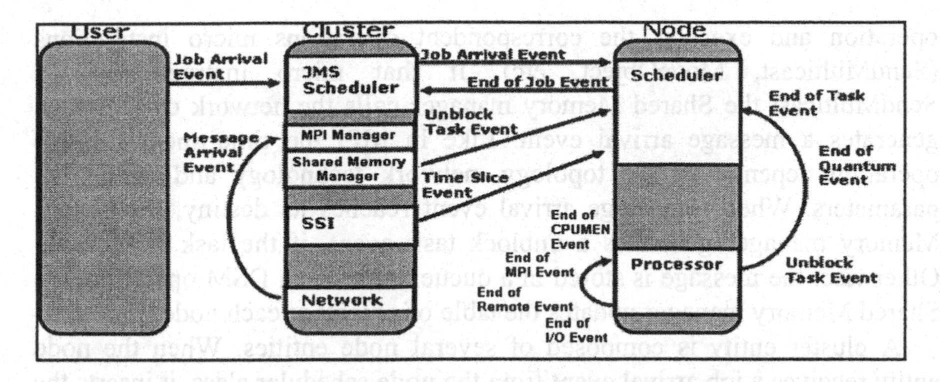

Figure 3. Events exchange diagram of ClusterSim

The Single System Image works as an operating system of the cluster, receiving and directing events to the responsible classes for the event treatment. Besides, it generates periodically the end of time slice event to indicate to the node schedulers that another time slice ended.

The network allows the cluster's nodes to communicate. Network modules include: physical topology, routing policies, message queues, network technology, and performance parameters. The current version offers two types of interconnection topologies: a bus and a switch. Both of them allow only point-to-point communication. The bus provides all nodes with a common medium through which two nodes are able to communicate at a given moment. The other nodes can read the state of the medium and are only allowed to transmit when it is not busy. In this model, no collisions were introduced. Thus, the bus allows only non-simultaneous point-to-point communication. The switch model allows pairs of nodes to communicate simultaneously. However, when a pair of nodes communicates, no other node can transmit or receive from them. Thus, the switch allows up to floor(N/2) simultaneous communications (N is the number of cluster nodes).

The network and MPI manager classes implement certain basic communication mechanisms, in order to support message-passing among tasks. As soon as a task executes an MPI function and the MPI manager recognizes it as a sending function, it calls the network class, which generates a message arrival event. The time spent for the message transmission from a node to another depends on the topology, network technology and contention parameters. When a message arrival event reaches its destination, the MPI manager generates a unblock task event, if the task is waiting. Otherwise, the message is stored into a queue.

The Shared Memory Manager classes were implemented for supporting shared memory with DSM write and read operations for objects distributed over a cluster simulated in ClusterSim. As soon as a task executes a DSM operation (Write, Read), the Shared Memory manager interprets the operation and executes the correspondent operations micro instructions (SendMulticast, MoveObject, etc). If that micro instruction is a SendMulticast the Shared Memory manager calls the network class, which generates a message arrival event. Like in MPI the time spent for this operation depends on the topology, network technology and contention parameters. When a message arrival event reaches its destiny, the Shared Memory manager generates a unblock task event, if the task is blocked. Otherwise, the message is stored in a queue. After each DSM operation, the Shared Memory manager updates the table of objects at each node.

A cluster entity is composed of several node entities. When the node entity receives a job arrival event from the node scheduler class, it inserts the new job's tasks into a queue. On each clock tick, the scheduler is called and executes the tasks using the processors of the node. An event is generated at the end of each CPU, I/O, DSM or MPI macro instruction within a task. A quantum is assigned to each task. When a task finishes, the processor generates an end of quantum event, so that the node scheduler executes the necessary actions (changing priority, removing the task from the queue etc). When the processor executes all the instructions of a task, an end of task event is generated for the node scheduler.

3.1.4 ClusterSim's Simulation Core

The simulation core is composed of the JSDESLib (Java Simple Discrete Event Simulation Library), a Java-based multithreaded discrete-event simulation library, developed by our group. Its main objective is to simplify the development of discrete-event simulation tools[1].

3.2 Simulation Model

A cluster is composed of: homogeneous or heterogeneous nodes, networks, job management systems, single system image, message-passing manager, and distributed shared memory. Each node has a memory hierarchy, I/O devices, one or more processors and an operating system. In our architecture model, a cluster is basically represented by a parallel job scheduler, a network, an MPI Manager, a Shared Memory Manager and a set of nodes.

$$Overhead = \left\lceil \frac{MessageSize}{MSS} \right\rceil \times ProtocolOverhead \qquad \text{(Equation 1)}$$

$$TransmissionTime = Latency + \frac{(MessageSize + Overhead) \times (1 + ErrorRate)}{Bandwidth \times 10^6} \qquad \text{(Equation 2)}$$

$$\mathbf{\textit{ElapsedTime}} = \mathbf{\textit{NumberOfInstructions}} \times \mathbf{\textit{CyclesPerInstruction}} \times \frac{1}{\mathbf{\textit{Frequency}}} \quad \text{(Equation 3)}$$

The scheduler implements parallel job scheduling algorithms, such as: space sharing, backfilling, gang scheduling etc. The network is represented by the following parameters: latency (ns), bandwidth (MB/s), protocol overhead (bytes), error rate (%), maximum segment size or MSS (bytes) and topology (Fig. 3). The equations 1 and 2 are used for calculating the message transmission time between two nodes. The MPI Manager is represented by MPI point-to-point primitives (blocking and non-blocking) and collective communication functions. And the Shared Memory Manager is represented by several consistency models and coherence protocols.

In our model, each node is represented by the following modules: scheduler, primary and secondary memory transfer rate and processors. The scheduler implements basic scheduling algorithms, such as: Round-Robin and FCFS. It also implements some algorithms that assure the correct operation of the cluster's parallel job scheduling algorithms. The primary memory transfer rate is used by the Shared Memory Manager and MPI manager, when receiving or sending data. The secondary memory transfer rate is used to calculate the time spent in the readings and writings of the I/O instructions. A node has one or more processors, and a processor is represented by a clock frequency and cycles per instruction rate (CPI). The processor uses the Equation 3 in order to calculate the elapsed time for executing n instructions of a program.

4. VERIFICATION OF THE NEW MODULES

ClusterSim v.1.0 was verified and validated in [1] and [13]. In this section, we verify the new simulation modules (MP and DSM), that were constructed on top of already validated modules belonging to ClusterSim v.1.0. In order to verify and test the new modules, we simulated two simple workloads and compared both simulation and analytical results.

The analytical results were obtained using real network values of a cluster (Fast Ethernet, Network Latency: 0.000179s, Data Payload: 1460 bytes, Network Bandwidth: 11.0516 MB/s, Protocol Overhead: 58 bytes, and Primary Memory Transfer Rate: 11.146 MB/s) and the equations 1, 2 and 3.

The first workload is a set of communication patterns that implement a broadcast operation with point-to-point primitives in a 16 processes job. We simulated the following patterns: Sequential Tree, Binary Tree, Binomial Tree and Chain[22,23]. As they have different parallel fractions, they perform differently over the switch model. Fig. 4 presents the analytical execution diagrams for the referred communication patterns. The diagrams represent the communications between processes in time (ms). A pair of cells with the same color in the same column (time interval) means that a pair of processes is communicating within that time interval.

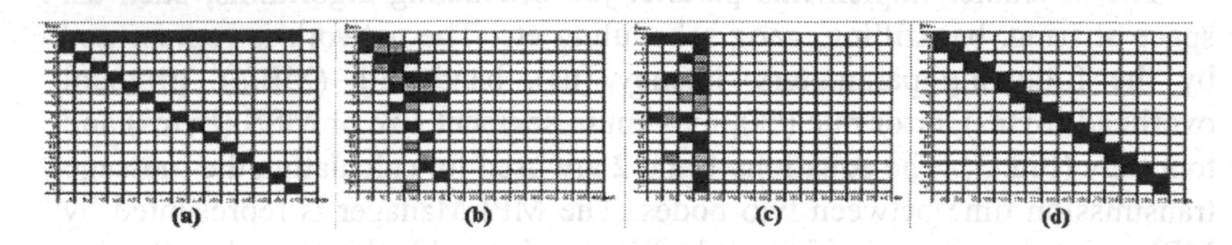

Figure 4. Analytical execution diagrams for the referred communication patterns: (a) Sequential Tree, (b) Binary Tree, (c) Binomial Tree and (d) Chain

In Fig. 4, we can see that Binomial Tree (c) yields the lowest response time among the analyzed patterns, because it best fits the switch topology of the network. Binary Tree (b) consumes more logical steps than the latter. Sequential Tree does not take advantage of the switch's parallelism because of its sequential nature. The Chain pattern presents the worst results over this network topology because it is sequential and because the dependency between its communications generates extra message transmission overhead.

Table 1. Response time (ms) for a 16 processes job over a switch topology

Comm. Pattern	Sequential Tree	Binary Tree	Binomial Tree	Chain
Res. Time (ms)	369.894301	147.971813	98.656046	369.977023

The results obtained in the simulator with 16 processes are presented in Table 1. They match the analytical results accurately (0.76% error), which means that the simulation was successful in this case. We made similar experiments with the reduction and the allgather operations for jobs having different sizes. They matched the analytical execution as well.

The second workload is a set of DSM operations that are managed by strong consistency models and some coherence protocols. First, in order to verify the DSM operations we constructed some jobs containing two tasks that execute DSM operations. Fig. 5 presents the analytical execution diagrams for the referred DSM operations. The diagrams represent the communications between tasks over time (ns). A remote operation is represented by a black line with the name of the operation. The dashed line represents a local DSM operation, which occurs when the node containing the calling task also contains the accessed object.

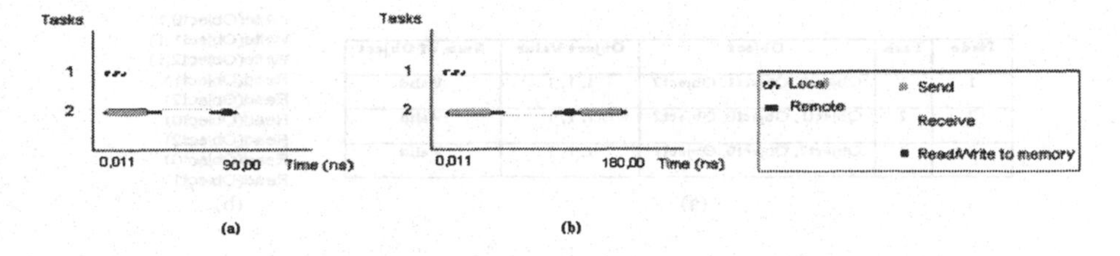

Figure 5. Analytical execution diagrams: (a) Remote Reads (b) Remote Writes

In Fig. 5, we can see that when a task has the object in its memory the DSM operation becomes local and the response time is lower than when the operation is remote. A remote write operation consumes more time than the local write, because a task has to obtain an object replica, write it, and send the updated object for all other tasks.

Table 2. Response time (ms) for a 2 tasks over a bus topology

Operation	Local Read	Remote Read	Local Write	Remote Write
Res. Time (ms)	0.011687	90.49662	0.011697	180.969828

The results obtained in the simulator with 2 tasks are presented in Table 2. They match the analytical results accurately (5.8% error), which means that the simulation was successful in this case.

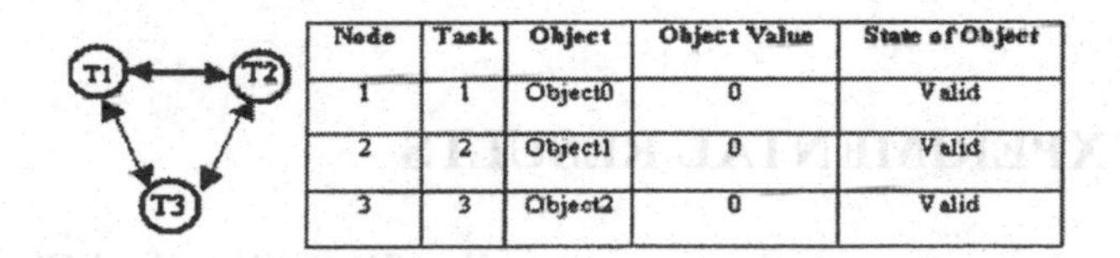

Node	Task	Object	Object Value	State of Object
1	1	Object0	0	Valid
2	2	Object1	0	Valid
3	3	Object2	0	Valid

Figure 6. Job used as the workload in Consistency Model verification

After that verification, we simulated DSM write and read operations on objects, using a sequential consistency model and the following coherence protocols: write-update and write-invalidate. Fig. 6 presents the job that was simulated for the verification of the sequential consistency model and the table that control the shared objects. In this job each task was located in one node and had one object. Each task has tree DSM operations: a Write to its object and two Reads to the objects of the other tasks.

Fig. 6 presents the original location and state of the objects (table of Objects and Nodes) at the beginning of the simulation. Fig. 7a presents the values in the table of Objects and Nodes after the execution of the Job with the sequential consistency model and the write update coherence protocol.

Node	Task	Object	Object Value	State of Object
1	1	Object0, Object1, Object2	1, 1, 1	Valid
2	2	Object1, Object0, Object2	1, 1, 1	Valid
3	3	Object2, Object0, Object1	1, 1, 1	Valid

(a)

Write(Object0,1)
Write(Object1,1)
Write(Object2,1)
Read(Object1)
Read(Object2)
Read(Object0)
Read(Object2)
Read(Object0)
Read(Object1)

(b)

Figure 7. (a) Table of Objects and Nodes after the job execution (b) Execution Sequence

In [24], this example was used to show a valid execution sequence for a job executed with sequential consistency model. In that example, a valid execution sequence could be: Write(Object0), Write(Object1), Write(Object2), Read(0), Read(2), Read(1), Read(2), Read(0) and Read(1). The resultant objects values for this sequence of execution is: 1, 1, 1, 1, 1, 1. As we could see in the Fig. 7b, our simulation got the same results as the analytical execution presented in [24].

Thus, we verified the new functions and parts of ClusterSim: MPI collective communication functions (with different communication patterns), DSM operations (write and read), consistency models, coherence protocols. The other functionalities and parts were not verified in this example. The consistency models, MPI collective functions, I/O instructions, network topologies and technologies had been verified exhaustively by means of more specific manual and analytical tests.

5. EXPERIMENTAL RESULTS

In this section, we present two case studies. We utilize the MP and DSM modules for cluster computing simulation. Our experimental environment is composed of two 16-node clusters interconnected by Fast Ethernet networks. One of the clusters used a bus interconnection device and the other used a switch. Each node had a Pentium III 1 Ghz (0.938 GHz) processor (0.9997105 CPI). The primary and secondary memories were unlimited and their transfer rates were respectively: 11.146 MB/s and 23.0 MB/s. The network had a 0.000179s latency, a 11.0516 MB/s bandwidth, a 1460 bytes segment size with a 58 bytes protocol overhead. Those values were obtained from benchmarks and performance libraries (Sandra 2003, PAPI 2.3 etc).

5.1 Message Passing Operations

In this case study, we created 10 workloads to test both clusters. The workload is composed of parallel jobs, containing different communication patterns. In order to simplify our experiments, we modeled only three MPI collective communication functions using different communication patterns[22,23]. We modeled: MPI_Bcast (one-to-all), MPI_Reduce (all-to-one) and MPI_Allgather (all-to-all). They are often used and are easy to understand and implement. MPI_Bcast and MPI_Reduce were implemented with the following patterns: Sequential, Binary and Binomial trees, and respectively with Chain and Ring. MPI_Allgather was implemented with: Star, FanIn-FanOut, Full FanOut, Circular, Shuffle and Pairwise[22,23]. All patterns were modeled as sets of point-to-point primitives and were used to transmit messages whose sizes were powers-of-two (from 1 Byte to 256KB). We will only present the results for 256KB messages, because the other sizes grew proportionally and their analysis would be repetitive.

In the bus based cluster, **MPI_Reduce** implemented with the Sequential Tree pattern presented the best results (see Table 3a). Ring presented the worst. As this model allows only one transmission at a time, it privileged the patterns with the smallest number of packets (combined messages). As a matter of fact, all patterns sent the same number of packets, but Ring presented an accumulated overhead due to waiting latency. Sequential Tree took advantage of the lack of memory limit. When a packet arrived, the receiving process consumed it as soon as a matching reception primitive was issued. Both Binary and Binomial presented the same results, which are slightly worse than those of Sequential (due to waiting latencies).

In the switch based simulated cluster, Sequential and Ring obtained the same results as in the bus based cluster. As their logics are sequential, they take no advantage of the switch's parallelism. On the other hand, Binomial presented the best results because the number of communicating processes doubled at each step. So this pattern took advantage of the parallelism of the switch, and took less logical steps than Binary.

In the bus based simulated cluster, three patterns presented the best results for **MPI_Bcast**. As shown in the Table 3b, Sequential, Binary and Binomial trees had the lowest response times for all number of processes and message sizes. They all spent exactly the same time to conclude, because they send N-1 packets and there is no dependency between packet transmissions. On the other hand, Chain presented the worst results due the same effect verified in the Ring reduction.

Over the switch, the dependency problem does not occur, as it does over the bus. So, both the Binary and Binomial trees yielded the same response times as in the reduction operation. The latter presented the best result, because it best explores the parallelism of the switch. Once again, Sequential Tree and Chain presented the same results as they did over the bus.

In the bus based cluster, Circular and Pairwise presented the best result over the bus for **MPI_Allgather** (see Table 3c). They are logically similar and they transmitted the smallest number of fixed sized packets, among all patterns. Shuffle presented a slightly worse performance than the latest couple, because the packet sizes doubled at each step. Besides both bus and switch allow no simultaneous bidirectional exchanges, which requires the steps to be divided in two sequential sub-steps. Over the switch, Pairwise presented the best result, because it takes advantages of parallelism with no bidirectional exchanges. In this case, Shuffle performed better than Circular.

As expected, the results showed that the best pattern for a given situation is the one whose logical topology best fits the physical network topology. Thus, the simulation tool provided the expected results coherently over the modeled bus and switch topologies. ClusterSim simplified time measurements because it didn't require any special techniques for measuring collective communications, so that they could be accurately measured.

Further experiments could also be possible if we considered: other communication patterns, different point-to-point primitives, other network topologies and technologies, buffer management, memory copies, packet routing, error rates etc. The simulation of some of these features would require the implementation of new modules for the simulator (e.g.: new network topologies). On the other hand some of the others can already be simulated without source-code changes.

Table 3. Response time (μs) of patterns on (a) reduction, (b) broadcast and (c) allgather

Model	Procs.	Seq. Tree	Bino. Tree	Bino. Tree	Ring
Bus	4	73,984,375	73,984,375	73,990,503	98,662,174
	8	172,622,038	172,628,166	172,622,038	197,324,347
	16	369,897,365	369,903,493	369,903,493	394,648,695
Switch	4	73,984,375	49,331,087	49,331,087	98,662,174
	8	172,622,038	98,649,918	73,996,630	197,324,347
	16	369,897,365	147,974,878	98,662,174	394,648,695

(a)

Model	Procs.	Seq. Tree	Bina. Tree	Bino. Tree	Chain
Bus	4	73,984,375	73,984,375	73,984,375	73,996,630
	8	172,622,038	172,622,038	172,622,038	172,658,804
	16	369,897,365	369,897,365	369,897,365	369,983,151
Switch	4	73,984,375	49,331,087	49,331,087	73,996,630
	8	172,622,038	98,649,918	73,996,630	172,658,804
	16	369,897,365	147,974,878	98,662,174	369,983,151

(b)

Model	Procs.	Star	F. FanOut	F.In F.Out	Circular	Pairwise	Shuffle
Bus	4	369,953,364	295,925,245	394,618,028	295,919,118	295,919,118	295,943,174
	8	1,553,745,293	1,380,939,542	1,701,744,227	1,380,933,414	1,380,933,414	1,381,068,560
	16	6,288,884,979	5,918,272,055	6,856,171,853	5,918,265,927	5,918,265,927	5,918,875,162
Switch	4	369,953,364	172,655,740	271,344,154	147,968,750	147,968,750	147,976,183
	8	1,553,745,293	789,199,348	1,011,258,315	394,569,036	345,250,204	345,306,203
	16	6,288,884,979	2,071,727,947	2,935,131,153	863,134,702	739,806,985	739,920,288

(c)

5.2 Distributed Shared Memory Operations

In order to simplify our experiments, we modeled only one consistency model and two coherence protocols with object replication in the nodes. The consistency model used in this simulation was sequential consistency, the strictest and most easily understandable consistency model. The modeled coherence protocols were: write-update and write-invalidate, which are often used and are easy to understand and implement.

In this case study, we created 4 workloads for testing over the bus based cluster. Each workload was a parallel job, containing different sequences of DSM operations for the shared object. Each one of the 4 workloads was a combination sequence of 4 DSM operations: LW (local write), RW (remote write), LR (local reads) and RR (remote reads). For each workload we varied the predominance level of remote and local operations between 20% and 80%. For example, a workload LW80 is a workload composed of 20% of local writes on shared objects and 80% of remote operations on shared objects. Thus, our 4 workloads were: LW80, RW80, LR80 and RR80. The DSM operations were modeled to execute operations on objects with powers-of-two sizes (from 1 Byte to 256KB). We will only present the results for 4KB objects, because the other sizes grew proportionally and their analysis would be repetitive. As we stated before in all simulations of DSM operations we use only the bus based cluster.

In order to simplify the visualization of the simulated results, in Fig.9, we used the letter U to indicate workloads executed with Write-Update coherence protocol and the letter I to indicate workloads executed with Write-Invalidate coherence protocol.

In Fig. 8a, we present the number of control and object messages for all workloads using both simulated coherence protocols. In Fig. 8b we present the job's response time for the simulated workloads. Considering the number of control messages sent (messages exchange between the node for controlling the remote operations), when we used the Write-Update protocol, LR and LW yielded the best results. In these workloads the remote shared objects were not frequently changed by the tasks, so the Shared Memory Manager didn't have to send many control messages to the nodes. For LR80, LW80 and RW80 workloads the write-invalidate presented the worst results (high number of control and object message), because on every change of a remote shared object its replicas were invalidated. So, whenever a node used an object after a subsequent write it had to obtain an object replica on that object. Only for the RR80 workload the number of messages was the same for both coherence protocols. That happened because in this workload only 20% of the operations were write operations. Also the objects messages presented the same results for both coherence protocols. Considering the job response time (Fig. 8b) we can see that the response time for the workloads executed with the sequential consistency model and the write-update coherence protocols yielded the smallest times. In this coherence protocol the number of exchanged messages was smaller, so the time wasted with network was smaller and the job's response time was better.

As expected, the results showed that the best results were obtained with the sequential consistency model using the write-update coherence protocol. Thus, the simulation tool provided the expected results coherently over the modeled protocols and model. Further experiments could also be possible if we considered: other coherence protocols, different consistency models, objects accessing protocols and other network topologies. The simulation of some of these features would require the implementation of new modules for the simulator (ex: new coherence protocols). That can be done by extending the existing coherence protocol classes. On the other hand some of the other coherence protocol classes can already be simulated without code changes.

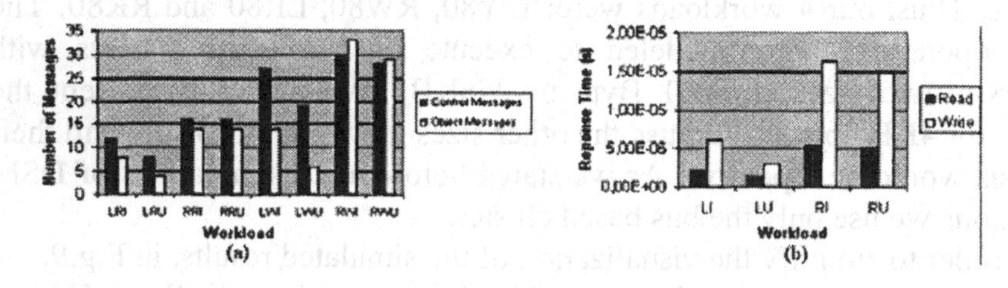

Figure 8. (a) Number of Messages x Workload (b) Response Time x Workload

6. CONCLUSIONS

In this paper we presented a new version of ClusterSim with the inclusion of Message-Passing (MP) and Distributed Shared Memory (DSM) modules. We verified the proposal and implementation of both modules by simulating: MPI collective communication functions using different communication patterns, DSM operations, consistency models and coherence protocols for object sharing. We also presented the new software architecture and simulation model. The main features of ClusterSim are: a hybrid workload model, a graphical environment, the modeling of heterogeneous clusters, and a statistical and performance module.

The graphical environment makes it easier to model, create and analyze the performance of clusters and workloads (parallel jobs and users). SRGSim and PP-MESS-SIM have detailed probabilistic workload simulation models that do not allow a detailed structural description of a job. Though the probabilistic description of a workload requires less simulation time and it is less representative than a structural description. On the other hand, the hybrid workload model of ClusterSim is composed of a probabilistic model and a structural description. It makes the simulation more deterministic than just probabilistic, thus making it feasible to represent real parallel jobs. Like Gridsim, ClusterSim supports heterogeneous nodes, but it also allows the configuration of interconnection networks with different parameters (ex:

topology, latency etc), as well as SRGSim and Simgrid. Like ClusterSim, MPI-SIM implements MPI collective communication functions with basis on point-to-point primitives. Nevertheless, the latter requires real C/C++/UC programs in order to run a simulation, whereas the former uses a hybrid language independent workload model. This model represents the abstract logic of the simulated program and is easier to understand. Besides, MPI-SIM only simulates point-to-point networks. Like DSMSim, ClusterSim also simulates distributed shared memory, but it provides more consistency models and coherence protocols, more statistical and performance information, and a graphical environment. Finally, ClusterSim has a statistical and performance module that provides data about the main metrics for the performance analysis of a cluster.

The main **contributions** of this paper are: the presentation, verification and analysis of MP and DSM modules in ClusterSim v.1.1. The new modules presented in this paper extended the functionalities of ClusterSim and enhanced its hybrid workload model. In the current version, it is feasible to model a workload by using instructions for: CPU/memory, I/O, point-to-point and collective communication, and DSM operations with memory coherence and consistency. MP and DSM were implemented on top of already validated modules and their verification was made by comparing analytical and experimental results. The verification led to a reliable simulation with very low error values. Thus, the simulator is yet a controllable and trustful environment with mechanisms for repeating and modifying parameters of real experiments.

The Java source code of ClusterSim is available and its classes are extensible. They allow the creation of new: network topologies, parallel job scheduling algorithms, collective communication functions and patterns, consistency models, coherence protocols, objects accessing protocols etc. As future works : implement a network topology editor and support to different routing algorithms, include new types of automatically generated workloads (ex: parallel programming models etc), include new memory consistency models and coherence protocols, simulate grid architectures, generate statistical and performance graphics, implement distributed simulation etc. More information at: http://ppgee.pucminas.br/lsdc_projetos.htm.

REFERENCES

1. L. F. W. Góes, Proposal and Development of a Reconfigurable Parallel Job Scheduling Algorithm. M. Sc. Thesis. Belo Horizonte, Brazil, (2004) (in Portuguese).
2. L. E. S. Ramos, L. F. W. Góes and C. A. P. S. Martins, Teaching And Learning Parallel Processing Through Performance Analysis Using Prober, 32nd ASEE/IEEE Frontiers in Education Conference, S2F13-18 (2002).

3. R. Buyya, and M. Murshed, GridSim: A Toolkit for the Modeling and Simulation of Distributed Resource Management and Scheduling for Grid Computing, The Journal of Concurrency and Computation: Practice and Experience, 14 (13-15), 1175-1220, Wiley Press (2002).
4. A.M. Law and W.D.Kelton, Simulation Modeling and Analysis, McGraw-Hill (1991).
5. Y.H. Low et al., Survey of Languages and Runtime Libraries for Parallel Discrete-Event Simulation, IEEE Computer Simulation, 170-186, (1999).
6. R. MacNab and F.W. Howell, Using Java for Discrete Event Simulation, 12th UK Performance Engineering Workshop, Edinburgh, 219-228, (1996).
7. H. Casanova, Simgrid: a Toolkit for the Simulation of Application Scheduling, 3rd IEEE/ACM International Symposium on Cluster Computing and the Grid, Los Angeles, 430-437, (2001).
8. H. Bodhanwala, et al., A General Purpose Discrete Event Simulator, Symposium on Performance Analysis of Computer and Telecommunication Systems, USA, (2001).
9. S. Prakash and R.L. Bagrodia, MPI-SIM: Using Parallel Simulation to Evaluate MPI Programs, Winter Simulation Conference (WSC98), 467-474, (1998).
10. J. Rexford, W.-C Feng, J.W. Dolter and K.G. Shin, PP-MESS-SIM: A Flexible and Extensible Simulator for Evaluating Multicomputer Networks, IEEE Trans. Parallel Distributed Systems, 8(1): 25-40, (1997).
11. D. Thaker and V. Chaudhary, DSMSim: A Distributed Shared Memory Simulator for Clusters of Symmetric Multi-Processors, International Conference on Parallel and Distributed Processing Techniques and Applications, 1561-1567, (2003).
12. L. F. W. Góes and C.A.P.S. Martins, RJSSim: A Reconfigurable Job Scheduling Simulator for Parallel Processing Learning, 33rd ASEE/IEEE Frontiers in Education Conference, Colorado, F3C3-8, (2003).
13. L. F. W. Góes, L.E.S. Ramos, C. A. P. S. Martins, ClusterSim: A Java-Based Parallel Discrete-Event Simulation Tool for Cluster Computing, IEEE Cluster 2004 Conference, (2004) (to appear).
14. Y. Zhang, H. Franke, E.J. Moreira and A. Sivasubramaniam, Improving Parallel Job Scheduling by Combining Gang Scheduling and Backfilling Techniques, IEEE International Parallel and Distributed Processing Symposium, 303-311, (2000).
15. B. B. Zhou, P. Mackerras, C. W. Johnson and D. Walsh, An Efficient Resource Allocation Scheme for Gang Scheduling, 1st IEEE Computer Society International Workshop on Cluster Computing, 187-194, (1999).
16. L. Breslau et al, Advances in Network Simulation, IEEE Computer, 33 (5), 59-67, (2000).
17. A. Sulistio, C.S. Yeo, and R. Buyya, Visual Modeler for Grid Modeling and Simulation Toolkit, Technical Report, GRIDS Lab, Dept. of Computer Science and Software Engineering, The University of Melbourne, Australia, (2002).
18. A. Sulistio, C.S. Yeo and R. Buyya, A Taxonomy of Computer-based Simulations and its Mapping to Parallel and Distributed Systems Simulation Tools, International Journal of Software: Practice and Experience, 34(7), 653-673, Wiley Press, (2004).
19. A Collection of Modeling and Simulation Resources on the Internet, URL: www.idsia.ch/%7Eandrea/sim/simindex.html
20. H.D. Schwetman, Using CSIM to Model Complex Systems, Winter Simulation Conference, (1988).
21. H-C. Hsiao and C-T. King, MICA: A Memory and Interconnect Simulation Environment for Cache-Based Architectures, IEEE 33rd Annual Simulation Symposium, 317-325, (2000).
22. W.B. Tan and P. Strazdins, The Analysis and Optimization of Collective Communications on a Beowulf Cluster, International Conference on Parallel and Distributed Systems, 659-666, (2002).
23. S.S. Vadhiyar, G.E. Fagg and J. Dongarra, Automatically Tuned Collective Communications, SuperComputing2000, 2000.
24. M. Dubois, C. Scheurich, and F. Briggs, "Synchronization, Coherence and Event Ordering in Multiprocessors", IEEE Computer, 21(2):9--21, (1988).

RENDERING COMPLEX SCENES ON CLUSTERS WITH LIMITED PRECOMPUTATION

Gilles Cadet, Sebastian Zambal, Bernard Lécussan
Supaero, Computer Science Department
10, av. Edouard Belin
31055 Toulouse Cedex
France
{gilles.cadet, sebastian.zambal, bernard.lecussan}@supaero.fr

Abstract This paper presents a parallel ray-tracing algorithm in order to compute very large models (more than 100 million triangles) with distributed computer architecture. On a single computer, the size of the used dataset generates an out of core computation. Cluster architectures designed with off-the-shelf components offer extended capacities which allow to keep the large dataset inside the aggregated main memories. Then, to achieve scalability of operational applications, the real challenge is to exploit efficiently the amount of available memory and computing power. Ray-tracing, for high quality image rendering, spawns non-coherent rays which generate irregular tasks difficult to distribute on such architectures. In this paper we present a cache mechanism for main memory management distributed on each parallel computer and we implement a load balancing solution based on an auto adaptive algorithm to distribute the computation efficiently.

Keywords: distributed ray tracing, cluster of commodity computers, load balancing, latency hiding

1. Introduction

Clusters are simple and widespread systems. Although scalability of operational applications remains the main goal, effective use of the distributed memory and the theoretical computing power becomes a difficult task when the number of computers increases. Scalability can be reached by hiding resource latencies and by using original mechanisms to tolerate residual latencies [Amdhal, 1967]. Furthermore, the system must also distribute computations and data while ensuring effective load balancing.

The distribution of the ray tracing algorithm requires a strong optimization of these various parameters. Moreover, dealing with large models which cannot be loaded completely into the main memory currently raises a technical challenge.

In the first part of this article, we present two methods to handle large datasets: a lazy ray tracing algorithm which limits the size of memory needed for the computations and a geometry cache management which gives the ability to access large dataset. In the second part, we expose a distributed solution of the ray tracing algorithm. We show an original method to approach the optimal load balancing and present ways of distributing the geometry database by duplication or by sharing. In the third part, we show the results obtained with the above methods. Lastly we discuss improvements of these methods that we are currently working on.

2. Background

2.1 Problems of Distributed Ray Tracing

The ray tracing algorithm is mostly used in 3D rendering to produce photorealistic images by accurate simulation of light propagation. To obtain efficient computation times, the simulation is done by following the inverse way of the light. Rays are spawned from the observer's eye through each pixel of the image to render and then are propagated inside the geometrical model by reflection and refraction laws until they reach light sources [Appel, 1968].

Using ray tracing to manage large dataset larger than main memory and to distribute the computations on commodity computers raises two types of problems. The first one is related to the use of distributed memory and the second one is due to the irregular data accesses caused by the traced rays.

Although the cluster architecture offers high cumulated computing power at low price, it suffers from low bandwidth and high latency of communication. Distributed memory implies the reorganization of standard algorithms in order to maximize the data access locality. To handle very large models, solutions exploit either the network and/or the local hard disk as a memory extension. These two points are strong constraints because the differences in bandwidth and latency between main memory and network or hard disk limit the performance of the global system.

From the algorithmic point of view, the rays hit the triangles of the model in a very irregular way. With a significant depth of recursion and a great number of secondary rays, the same triangles are requested several times in an unforeseeable order.

To summarize, the various problems we have to consider are: efficient use of distributed memory, low bandwidth and high latency of remote memories and irregularity of data accesses. Below we give an overview of recent works that treat these problems.

2.2 Previous Work

To reduce rendering time current studies develop three approaches. One of the older solutions is parallel computing either with parallel machines or clusters. The second approach relates to the capacity of new graphic cards to load custom code [Purcell and al, 2002, Purcell and al, 2003] the use of these graphic cards is quite limited because they are only optimized for Z-buffer rendering. However, more recent proposals like AR350 [Hall, 2001] and SaarCor [Schmittler and al, 2002] use graphic cards with a built in ray engine. In fact fields of research related to graphic card solutions are very promising with regard to performances, but are still limited to models which fit in the graphic card memory or in main memory.

Work on large models with clusters is justified overall by the need to visualize the models realistically as a whole and not by parts. Compared to other algorithms like Z-buffer, ray tracing has a complexity which grows logarithmically according to the number of triangles contained in the model. This logarithmic property is guaranteed only if the model can be loaded completely into memory.

Ray tracing can be distributed naturally by computing independently groups of rays on each computer. The master splits the image into parts distributed to each slave in a demand driven computation. In this case, the geometry needed by each slave is transmitted via the network or can be locally loaded from the hard disk. Another way of distributing is to share the model between all computers; the distribution is then controlled by the geometry owned by each slave (data driven) and the rays are propagated among the machines. A third and last way of distributing is a hybrid solution which is a combination of the two preceding methods. More recent works use demand driven distribution because it minimizes exchange of information over the network improving the load balancing. So we expose below some recent results.

The RTRT team [Wald and al, 2001a] proposes the first animation of large models by distributing an optimized ray tracing engine on commodity computers [Wald and al, 2001b]. The model is precomputed by the construction of a binary space partition (BSP) tree. Each voxel of this tree is an independent element recorded in a file which contains geometry, textures, precomputed data and a BSP subtree. One voxel has an average size of 256 KB and is directly loadable into main memory.

To take advantage of the bi-processor architecture, each slave runs two threads for processing rays. Moreover there is another thread for loading the data asynchronously. The geometry is centralized on a data server which distributes the voxels when the slaves request them. Their downloadable animation shows a powerplant of 12.5 million of triangles with outdoor and indoor views. The file has a size of 2.5 GB after a precomputation of 2 hours and 30 minutes. Results perform about 5 frames per second on a cluster of 7 computers. The animation requires an average data bandwidth of 2 MB/s with some peaks about 13 MB/s which saturate the network. The analyse of these results seems to show that the working set stays rather small which indicates that impacted triangles are localized and that the slaves do not often have to deal with low bandwidth and high latency of the network.

Another study [Demarle and al, 2003], which is very close to the RTRT's one, but less optimized in the ray core engine, shows another possible distribution on a cluster to handle large dataset. The main difference is the way of recovering data by each slave. Each one launches two processing threads and a voxel loader thread, called "DataServer", which emulates a shared memory. A processing thread sends its geometry data requests to the local DataServer which either loads the data from the main memory or transmits the received request to a remote DataServer. To work efficiently, the model is partially loaded by each slave and thus is wholly loaded into the distributed memories. The performance reaches more than 2 frames per second for a model size of 7.5 GB with a cluster of 31 processors. This model must be precomputed before the rendering using hierarchical voxels in 40 minutes.

2.3 Goals

As we have mentioned above, current studies render very large scenes on commodity computers efficiently. All these studies render images in "real time" after a precomputation step which gives the possibility to load parts of the model independently and with spatial coherency. Each slave manages a voxel cache with a least recently used (LRU) policy. However, the suggested methods are only effective for a quite small working set and moreover after a time consuming precomputation step.

The study presented in this paper wants to differ in this last point by strongly reducing the precomputation step. Using a traditional model file we seek to optimize the total rendering time, taking care of the complete computation from the model recording to its visualization. This approach makes sense with the modelling process which unceasingly changes the geometry integrity. By reducing the precomputation time, we expect that the time of the first visualization can be considerably reduced.

Moreover, we want to handle very large models while keeping a photorealistic quality of the rendered image without any restrictions (e.g. without limiting the depth of recursion of the propagated rays). The objective is to render models which include about 100 million triangles.

3. Methods to Handle Large Dataset

To handle large dataset with ray tracing, the strategy of finding and loading data must be modified. We first describe a method to quickly find data without using extra memory and then we present a cache manager which offers the possibility of working with a model that is larger than the main memory.

3.1 Lazy Construction of the Octree

A geometrical model is built with objects which are discretized into triangles. Each triangle is linked to physical properties giving its color and the way to reflect and refract rays when they hit its surface [Whitted, 1979]. The hot spot when developing a ray tracing engine is to have an efficient function to compute the intersection between triangles and rays.

To reach this goal, the model is generally divided into voxels by a three-dimensional regular grid. However, with voxels of the same size, the distribution of the triangles is unbalanced; this introduces useless computation time to traverse empty voxels. Moreover, this structure takes a lot of memory because the dimension of the grid must be large enough to keep the number of triangles in denser voxels as small as possible. A well known solution is to build a hierarchy of voxels like an octree [Glassner, 1984]; in this case only not empty voxels are divided into subvoxels which saves memory. However, the exploration of the octree is a little more complex because of the hierarchy and the irregularity of the data structure.

The dynamic construction of the octree is an efficient solution to keep the memory occupancy low [Bermes and al, 1999]. Instead of building the whole octree, the algorithm starts with only one voxel which represents the bounding box of the model. During the rendering computation, each time a ray traverses a voxel not yet divided, the subvoxels are dynamically built and the traversal of the children is done recursively; with this principle, only impacted triangles are loaded into memory and are used directly. This construction, called lazy octree construction, is the basic technique to quickly find triangles in large models.

This lazy algorithm shows the following properties: first, a child node is evaluated only if it contains necessary data for the computation; then, the node evaluation result is definitively stored in the octree and will be reused for neighbor ray computation. With distributed memory architecture, the data structure is built dynamically and locally for the subset of rays computed by each processor. Thereby, the algorithm exploits spatial ray coherence.

Efficiency is obtained for models which fit in the main memory of one computer (see results in Table 2). The main objective of this new study is to implement mechanisms to overcome this limitation.

3.2 Geometry Cache Management

To get along with the problem of huge dataset, in a first step the large model is subdivided into small blocks. These blocks are loaded into a cache when they are requested by the ray tracing engine (Figure 1). It must be taken care that this geometrical cache does not interfere with the virtual memory of the operating system. Recent works have demonstrated that classical virtual memory is not very effective for applications using large quantities of data [Cozette, 2003]. Overall there are two ways to deal with this: either the virtual memory of the operating system is adapted or the flow of data is controlled directly. In order to have free control over the caching process we have decided to implement the later solution.

Before considering parallelization, a sequential version of the cache was developed. One part of the complete caching system is a compiler which initially converts the model into a structured binary file. This binary file then serves as a database for the actual cache supplying the ray tracing engine with blocks.

The geometrical data of a model is initially given as a text file containing triangles. The compiler transforms this text file into a binary file which consists of a sequence of blocks. Each block contains information about several triangles and can be loaded into memory independently. To accelerate the system, a non destructive LZO compression [Oberhumer, 2002] is applied to the blocks which reduces the size of the blocks by about a half and leads to a speed-up of 25% for the time spent for loading the model. A well chosen size for the blocks minimizes the redundancy and maximizes the compression rate which leads to better performance. Experimentally we found an optimal block size of 64 KB.

The compilation of the model is the only needed step to initialize the cache memory and to start the image rendering; this takes about three minutes for a model size of 3.4 GB.

During the rendering process the cache delivers requested triangles to the ray tracing engine. Each time a triangle is demanded, the cache first has to determine the block containing the triangle; if the block is not in memory at this time the cache has to load it. That might include unloading another block. Finally the demanded triangle can be returned.

The cache management is purely associative. A table indicates for each block if it is loaded or not. Thus the time to find a block is negligible and the most important item to improve the cache efficiency is to increase the hit rate.

The cache policy of unloading blocks is LRU. The number of allocations (and deallocations) of memory is minimized at each cache miss by considering memory areas that have been allocated before. This approach does not fragment the memory and limits the number of system calls.

Figure 1. Data flow of the geometry cache.

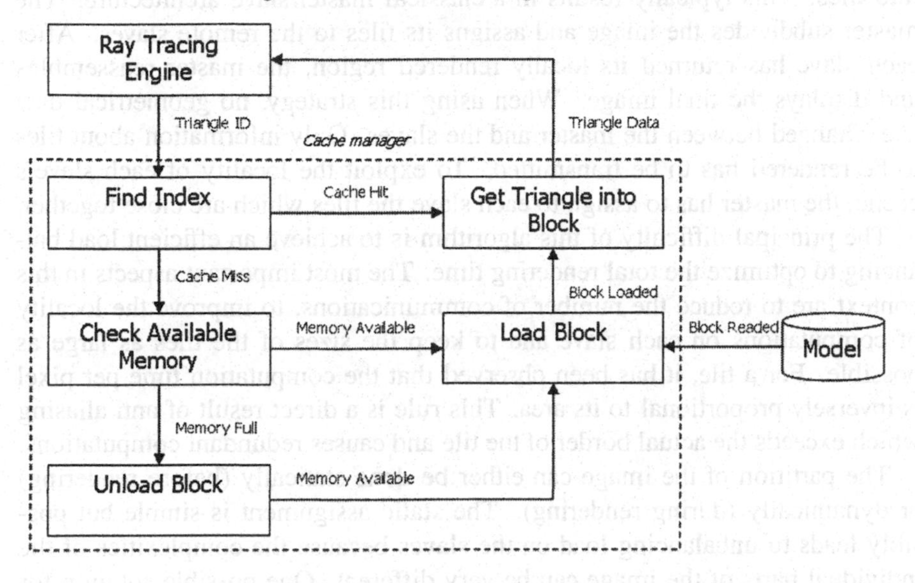

We have tested several strategies to reduce the time of loading blocks and also to proceed the computations while loading blocks. The first idea was to use two different levels of subdivision: large blocks for loading data and small blocks for internal use in the cache. Another idea was to use a small secondary cache which is able to load a sequence of blocks simultaneously. Furthermore we tried to apply an own thread dedicated only to anticipate block loading. However all these approaches did not lead to notable improvements of the rendering time. In fact, all the overhead caused by the additional complexity (useless loading of blocks, management of additional functionality, overhead when synchronizing threads) lowered the performance of the global system.

4. Distribution of Large Dataset

What it has not been answered so far is the question how to handle large geometrical data in a distributed environment. The following topics discuss the task distribution as well the data distribution.

4.1 Task Distribution

In this section we will first talk about how to distribute computations for the ray tracing algorithm in general. Then we will present an efficient method to dynamically reassigning the work load of the slaves.

The rendering process is subdivided according to a partition of the image into tiles. This typically results in a classical master/slave architecture. The master subdivides the image and assigns its tiles to the remote slaves. After each slave has returned its locally rendered region, the master reassembles and displays the final image. When using this strategy, no geometrical data is exchanged between the master and the slaves. Only information about tiles to be rendered has to be transmitted. To exploit the locality of each slave's cache, the master has to assign to each slave the tiles which are close together.

The principal difficulty of this algorithm is to achieve an efficient load balancing to optimize the total rendering time. The most important aspects in this context are to reduce the number of communications, to improve the locality of computations on each slave and to keep the sizes of the tiles as large as possible. For a tile, it has been observed that the computation time per pixel is inversely proportional to its area. This rule is a direct result of anti aliasing which exceeds the actual border of the tile and causes redundant computations.

The partition of the image can either be done statically (before rendering) or dynamically (during rendering). The static assignment is simple but possibly leads to unbalancing load on the slaves because the complexities of the individual parts of the image can be very different. One possible solution for this problem is to estimate the complexities in a preprocessing step. However this additional preprocessing costs a lot of time and moreover the estimation of complexities might not be sufficient to achieve a good load balancing [Farizon and al, 1996]. The conclusion is that a dynamic subdivision of the tiles has to be applied.

To subdivide the image, the size of the tiles has to be chosen carefully. This size is guided by the following antagonism: large tiles reduce the number of communications and increase the time per pixel; small tiles allow a fine regulation of the load balancing which is particularly important at the end of the rendering process. It is thus interesting to subdivide the image into large tiles at the beginning of the rendering and then to subdivide them before the rendering ends. The difficulty is to know when exactly the subdivision should start.

To find a good solution for the subdivision, three main items are considered:

- the definition of central points which guide the assignment process,
- the assignment process itself,
- an adaptive subdivision method for the assigned tiles.

Definition of the Rendering Zones. The initial granularity of the assigned tiles is chosen in a way to achieve at least five tiles per slave. This value has empirically been found out to satisfy the models used.

To preserve the locality, we assign to each slave a position in the image plane called central point. The tiles assigned to each slave are as near as possible to the slave's central point. Two different ways to choose the central points were examined. The first approach was to place the points into the image in form of an X (Figure 2. left); this strategy assumes that most images have their highest complexity in the center and allows treatment of the largest tiles there. The second approach places the central points homogeneously taking into account the image resolution (Figure 2. right); this generally leads to better performances.

Figure 2. Tile assignments with 6 slaves. Each slave has a central point and is represented by a specific colour. Left: "X" placement of central points and nearest tile policy to assign the next tile. Right: homogeneous placement of central points and trade off policy to assign the next tile.

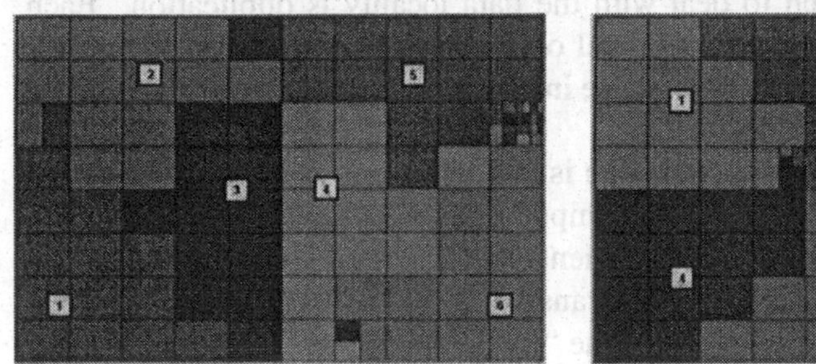

Assignment and Subdivision. The assignment of the tiles must reuse already built parts of the octree by optimizing the data locality. One strategy we tried out was to assign the free tiles nearest to each slave's central point. Although this method was efficient at the beginning of the rendering there may be no free tile near a central point at the end. A slightly modified strategy keeps the tiles near the central points unassigned as long as possible. The tiles are now assigned by minimizing the distance to the own central point and maximizing the distances to the other central points.

Large tiles are defined at the beginning of the rendering and are subdivided during the rendering process. An algorithm using the estimation of the remaining computation time allows us to subdivide the tiles as late as possible and leads to an efficient load balancing. An estimated remaining computation time is associated with each free tile as a function of previously rendered tiles. Each estimation is weighted according to the inverse of the distance between the currently and the previously rendered tiles. When a new tile has to be assigned, the master checks if the total time of the current and future computations is greater than the time estimated for this tile. If this is not the case, the tile is subdivided and the tile assignment function is recalled. This continues until an acceptable computation time or a minimum size of the tile is reached.

The principle of image subdivision with the assignment and the subdivision of tiles offers an efficient load balancing (Table 4). The method of estimating the remaining time in fact suffers from underestimation but keeps the scale between estimated intervals and real intervals. A little amelioration of load balancing could be realized by assigning an amount of time to each tile. When this critical time is exceeded, the slave must stop and retransmit the partially computed tile.

5. Data Distribution

5.1 Duplication

The simplest approach to deal with the data locality is duplication. Each slave has a complete copy of the model on its hard disk. Consequently, each slave is able to render a part of the image independently of the others, accessing its local data using a local cache.

The main advantage of this approach is that almost no communication over the network is necessary during the computation. The master only communicates with the slaves to send all assignments and receive all rendered tiles. The main drawback is the time needed to transfer the file to each local disk. This transfer takes about three minutes for the "Engine x729" model (Figure 5) and can be partially masked by the time needed to parameterize the computation by the user.

5.2 Sharing

When a slave needs a block not yet loaded locally, it can avoid the loading of the block from the hard disk if this block was already loaded into the main memory of another slave. We will now discuss the conditions under which it is better to access remote caches for a requested block than to load it from the local disk.

To find the more effective strategy (either to load the block or to transmit it over the network) we have compared the communication bandwidth between two slaves using MPI (Message Passing Interface) to the hard disk bandwidth. With a Gigabit Ethernet network the bandwidth of MPI is better for block sizes greater or equal to 16 KB. The bare latency of MPI communication was measured with about 136 μs compared to 5000 μs for the hard disk (for random access).

According to the above characteristics the previous cache mechanism has been modified. The function that loads the data from the disk is replaced by a function that downloads the requested block from another remote cache.

Two different kinds of slaves are used. The first plays the classical role of a "worker" and uses a cache with limited memory. If the worker has to load a new block, it transmits its request via MPI to another slave. The other slave receiving the demand is in the role of the "reader". This type of slave only waits for the block requests and does not communicate with the master. If the requested block is available the reader simply returns it to the worker.

The tests, carried out with a small model but with a high miss rate of the cache, show that the data accesses over the network allows to double the rendering performance. Because the network offers better performance than the hard disk with models that cause a high miss rate of the cache, it is interesting to study an architecture that uses remote accesses.

We will consider two possible configurations. The first is to use as many readers and as many writers as there are machines available. With this configuration, there is a reader for each worker but the memory is used commonly by all slaves. The second configuration is to use a single reader. This means that a single machine is busy with loading the blocks and distributing them over the network to the other slaves.

During its initialization phase, each slave receives a role assignment from the master. If it is a worker, it waits for the tile assignments. If the slave is a reader, it links itself to the local files of blocks and waits for workers' requests.

Each worker is related to all the readers and each reader manages a part of the geometry. The geometry is shared uniformly so that each reader has to manage the same amount of data. A worker sends its requests to a specific reader depending on the index block to load.

This algorithm is applicable for models that have a file size smaller than the total readers' memory. In fact access over the network is not efficient when blocks have to be loaded from the remote hard disk. Each used block should reside in one of all memories to avoid remote hard disk accesses.

Table 1. Rendering times depending on the number of readers.

Medium for loading	Hard Disk	Network	
Nb. of Readers	-	1	4
Nb. of Workers	4	4	4
Rendering Time (sec.)	70	37	35

The above table (Table 1) shows the execution times depending of the roles given to the slaves. The "Engine" model (Figure 5) used for this measurement has a size of 5 MB with blocks of 4 KB. Furthermore the memory of the workers is limited to 10% of the total model size. The readers can load the whole model into the cache. These results show that it is much more efficient to load the blocks over the network than to load them from the local hard disk. The best result is achieved by using 4 readers because this avoids a network bottleneck.

6. Results

The results are obtained with a cluster of six machines linked together by an Ethernet network with a bandwidth of 1 GB/s. Each machine is equipped with an AMD Athlon MP 2000+ biprocessor (1.7 Ghz), has got an IBM Deskstar 60 GXP hard disk (60 GB capacity, 7200 rpm, 2 MB of cache) and 1 GB of DDRAM. The operating system used is Windows 2000 Professional.

6.1 Lazy Octree Construction

The following performance comparison (Table 2) shows the efficiency of the lazy octree construction in terms of computation time and memory size needed. Previous comparisons can be found in [Bermes and al, 1999].

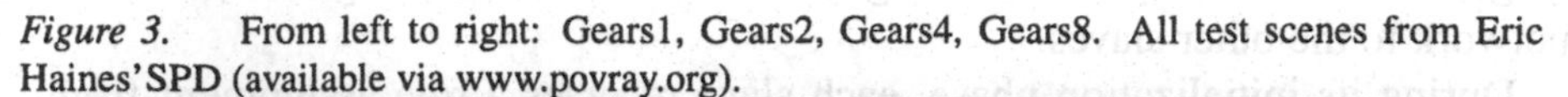

Figure 3. From left to right: Gears1, Gears2, Gears4, Gears8. All test scenes from Eric Haines' SPD (available via www.povray.org).

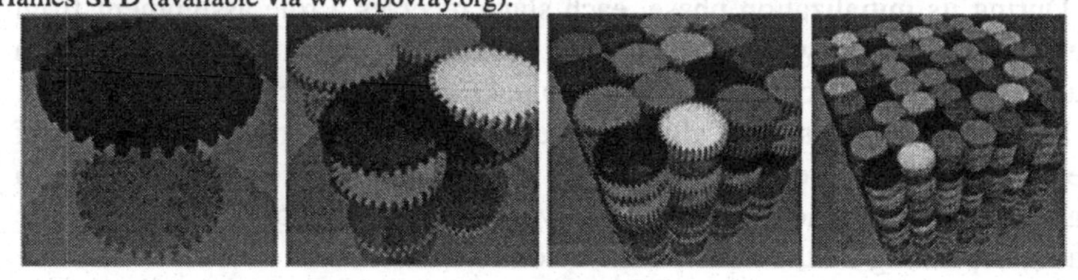

All rendered images (Figure 3) have a size of 512x512 pixels and are computed by only one machine. PovRay for Windows V3.5 is used and configured with its default options except for the following parameters: deactivation of the anti-aliasing, first pass with ray tracing and shadowing with ray tracing.

Table 2. Performance comparison between PovRay and ray tracing with lazy octree construction.

Model	Gears1		Gears2		Gears4		Gears8	
Number of triangles	27 552		219 744		1 757 280		14 057 568	
Ray Engine	Pov	Lazy	Pov	Lazy	Pov	Lazy	Pov	Lazy
Parse Time (sec.)	0.0	0.1	1.0	0.3	3.0	2.3	153.0	15.8
Trace Time (sec.)	17.0	7.8	16.0	6.9	17.0	6.7	21.0	7.6
Full Time (sec.)	17.0	7.9	17.0	7.2	20.0	9.0	174.0	23.4
Peak Memory (MB)	0.6	0.7	4.1	1.9	32.3	9.3	258.6	58.9

Theses results confirm that the lazy octree construction can efficiently save memory, especially for large models, and reduce the total computation time.

6.2 Cache Management Performance

The cache management is tested with a large model called "Engine x729". This model is built using 83 million triangles contained in a file of 3.4 GB (not compressed) with approximately 40 bytes per triangle. The following table (Table 3) summarizes the various tests. After compression, the blocks' file takes 1.8 GB on the hard disk with block size of 64 KB. Measurements are made on only one machine. Evaluation of the chosen model faces up to difficulties because it includes a lot of transparencies which implies that the major part of the triangles is hit. Results show that a model of more than 83 million triangles can be rendered with only one machine in less than one hour.

Table 3. Geometry cache performance with one machine.

Ray Depth	1	2	12
Rendering Time (RT)	59.4 min.	82.7 min.	421 min.
Loading Time (LT)	31.3 min.	47.5 min.	-
Nb. of Rays	1.20 e+06	1.56 e+06	-
Nb. of Cache Accesses	1.86 e+09	1.93 e+09	-
Nb. of Block Loads	1.26 e+06	1.50 e+06	-
$RT/(RT-LT)$	2.11	2.40	-

The next figure (Figure 4.) shows that the size of the octree increases constantly during the computation from 320 MB to 625 MB. This means that the cache manager runs with only 200 MB because it adapts dynamically to the free available memory. Thus the cache management works, at the end of computations, with less than 6% of memory compared to the model size. This small part of memory does not disturb the progression of computations because the increasing accuracy of the octree during the rendering process permits the localization of the ray engine data requests.

6.3 Ray Tracing Parallelization

Table 4. Load balancing performance with temporal refinement of the tiles (Times in seconds). The parallel efficiency is: $\frac{SequentialTime}{ParallelTime \times NumberOfSlaves}$.

		Sequential	2 slaves		4 slaves		6 slaves	
Model	Nb. Tri.	Time	Time	Eff.	Time	Eff.	Time	Eff.
Engine	114 7977	31.7	17	93%	9.6	83%	7.1	75%
Stairs	74 182	76.2	39.3	97%	21.3	89%	15.3	83%
ChevyInd	29 630	170.2	90.3	94%	49.1	87%	35	81%

Figure 4. Memory evolution while rendering.

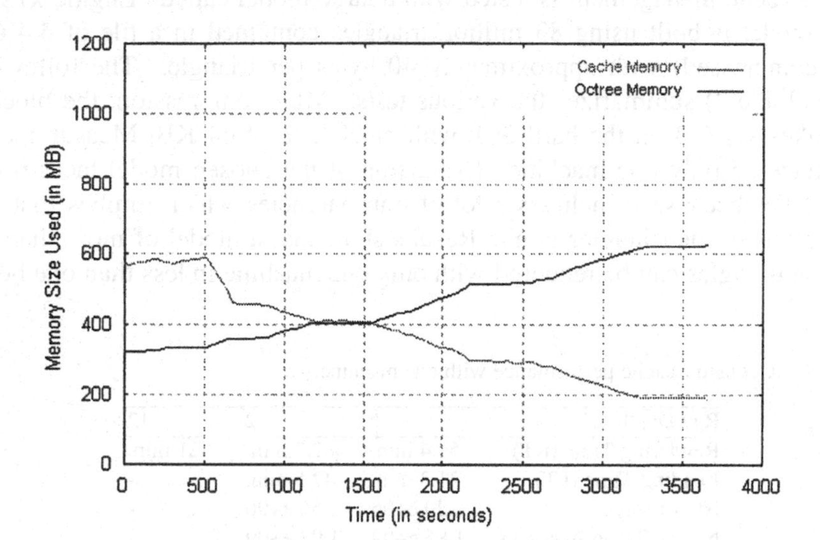

The above table (Table 4) summarizes parallelization benches done with one, two, four and six slaves for three different models. All measured times are obtained with the temporal refinement method with a single parameter setting:

- the initial size of the tiles is set to 64x64 pixels,
- the minimal size of the tiles is limited to 8x8 pixels,
- an option allowing the slave to stop an assigned tile after a critical time is used.

Thanks to this adaptive core, which does not need specific parameters for each model, we obtain an efficient load balancing. The parallel efficiency with 6 slaves is close to 80% and thus offers a good acceleration of the rendering time.

For comparison, we keep the principle of dynamic affectation of the tiles but we inhibit the subdivision procedure. Like in another study [Wald and al, 2001b], a fixed size of the tiles (32x32 pixels) is set. Results show that parallel efficiency decreases about by 3% and thus validate the contribution of the temporal subdivision method of tiles.

Moreover the adaptive method presented easily deals with the problem of heterogeneous computers [Qureshi, 2000] because it balances the computation load according to the power of each computer.

6.4 Large Dataset Distribution

The following tests have been carried out with using the cache management and the parallelization of the ray tracing together. Each slave has a local copy of the model on its hard disk.

Figure 5. Left: rendered image of the "Engine" model (5 MB, 114.000 triangles). Right: rendered image of the "Engine x729" model (3.4 GB, 83 million triangles).

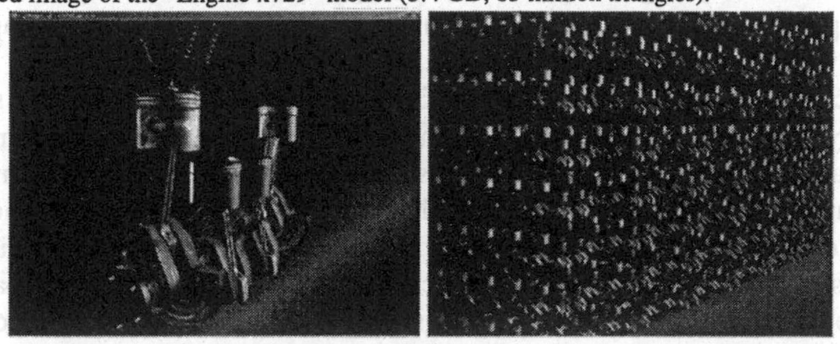

The "Engine x729" model (Figure 5) is rendered in 74 minutes with a parallel efficiency of 95% for a depth of 12 (Table 5) which corresponds to more than 13 million rays traced. This efficiency must be carefully interpreted because during the sequential execution the octree size exceeded 900 MB and the cache manager reached a minimal size of 100 MB which caused a memory system swap. But in parallel, thanks to the distributed memory each slave can efficiently finish its execution.

Table 5. Geometry cache performance with 6 machines. The last column shows the ratios of the distributed computations to the sequential computation.

				Ratios
Ray Depth	1	6	12	-
Rendering Time (RT)	26 min.	51 min.	74 min.	-
Loading Time (LT)	12 min.	27 min.	44 min.	-
Nb. of Rays	1.19 e+06	8.40 e+06	13.8 e+06	1.0
Nb. of Cache Accesses	5.51 e+09	7.08 e+09	7.64 e+09	2.9
Nb. of Block Loads	3.50 e+06	5.20 e+06	6.78 e+06	2.8
$RT/(RT-LT)$	1.9	2.1	2.4	-
Parallel Efficiency	38%	-	95%	-

With a ray depth of 1, the parallel efficiency is only 38%. This lack of scalability is due to an high memory miss rate within each slave. To obtain good performances, the ratio values of the last column should be near to one meaning that each slave can work in a complementary way. This overloading comes from a trade off between the depth of the octree and the number of triangles inside each voxel. Actually, the leaves of the octree contain too many triangles which implies an increase of data requests on such large models.

7. Conclusion

The computing power and the size of available memories of a cluster allow to handle very large datasets and to compute complex models efficiently. In this paper we have presented solutions and mechanisms to render complex scenes without using a time consuming model precomputation. First, a lazy ray-tracing has been implemented to limit the precomputation time and to save memory. Then, we have used each machine to cache in its own main memory the dataset stored locally on the hard disk and to compute parts of the image distributed by an original master/slave algorithm. Finally, as the whole model can be distributed within all main memories we have developed mechanisms to access required data remotely taking care of the high latency and low bandwidth of the network. Presented results show the efficiencies of the proposed solutions with some drawbacks that we expect to solve within future works.

8. Future Work

Related experiences show that it is important that the finer granularity of the octree does not cause extended use of memory (Figure 4). It seems possible to reduce the size of references from voxels to triangles and from triangles to vertices by limiting the scope of these references to each voxel. As an example, the maximum number of triangles inside a voxel is 65 536 which can limit the reference size to two bytes instead of four actually. This kind of optimizations already allows to increase the exploration depth of the octree.

To increase the locality of computations and to minimize the cache misses, the blocks of data must be made spatially coherent by a preprocessing step of the model. To reduce the execution time of this preprocessing step, we suggest to use two levels of granularity for the octree management. The coarse level, which is precomputed, is sufficient to create blocks of data that can be loaded into memory directly (about 64 KB). The fine level of the octree, which is used for efficiently finding intersections between rays and triangles, is built during computations with a lazy algorithm.

Using voxels as blocks allows an implementation of an asynchronous process for loading blocks. According to Matt Pharr [Pharr and al, 1997] a buffer of rays could be used to obtain the ability to treat rays while loading blocks. Furthermore it will be possible to examine various strategies for selecting rays from this buffer to increase locality and to improve performance.

With these mechanisms we expect to improve scalability when using more machines. The performance will allow us to realize animation in "real time" and to face new topics like dynamic moving of objects [Reinhard and al, 2000, Lext and al, 2001] or the elimination of temporal artifacts [William and al, 2001].

Acknowledgments

The authors want to thank Christophe Coustet and Sébastien Bermes for their invaluable councils on the techniques for wave propagation simulation and especially photorealistic rendering.

References

[Amdhal, 1967] G. Amdhal. *Validity of the single-processor approach to achieving large scale computing capabilities.* In AFIPS Conference Proceedings vol. 30 (Atlantic City, N. J., Apr. 18Ű20). AFIPS Press, Reston, Va., 1967, pp. 483-485.

[Appel, 1968] A. Appel. *Some techniques for shading machine renderings of solids.* In Proc. Spring Joint Computer Conference (Atlantic City, April 30-May 2, 1968), AFIPS Press, Arlington, Va., pp. 37-45.

[Bermes and al, 1999] S. Bermes, B. Lécussan, C. Coustet. *MaRT : Lazy Evaluation for Parallel Ray tracing.* High Performance Cluster Computing, Vol 2 Prentice Hall 1999.

[Cozette, 2003] O. Cozette. *Contributions systèmes pour le traitement de grandes masses de données sur grappes* Thèse de l'Université de Picardie Jules Verne, Amiens, soutenue le 18 décembre 2003.

[Demarle and al, 2003] D. E. Demarle, S. Parker, M. Hartner, C. Gribble, C. Hansen. *Distributed Interactive Ray Tracing for Large Volume Visualization.* IEEE Symposium on Parallel and Large-Data Visualization and Graphics October 20 - 21, 2003 Seattle, Washington.

[Farizon and al, 1996] B. Farizon, A. Ital. *Dynamic Data Management for Parallel Ray Tracing.* Computer Science Department, Technion, Haifa, Israel Ű Mars 1996.

[Glassner, 1984] A. Glassner. *Space subdivision for fast ray tracing.* IEEE Computer Graphics and Applications. 4(10), pages 15-22, 1984.

[Hall, 2001] D. Hall. *The AR350: Today's ray trace rendering processor.* In Proceedings of the Eurographics/SIGGRAPH workshop on Graphics hardware, Hot 3D Session 1, 2001.

[Lext and al, 2001] J. Lext, T. Akenine-Möller. *Towards rapid reconstruction for animated ray tracing.* In Eurographics 2001 Ű Short Presentations, pages pp. 311Ű318, 2001.

[Oberhumer, 2002] M. F. X. J. Oberhumer. *http://www.oberhumer.com/opensource/lzo/.*

[Pharr and al, 1997] M. Pharr, C. Kolb, R. Gershbein, P. Hanrahan. *Rendering Complex Scenes with Memory-Coherent Ray Tracing.* Proceedings of SIGGRAPH 1997.

[Purcell and al, 2002] T. J. Purcell, I. Buck, W. R. Mark, P. Hanrahan. *Ray Tracing on Programmable Graphics Hardware.* In to appear in Proc. SIGGRAPH, 2002.

[Purcell and al, 2003] T. J. Purcell, C. Donner, M. Cammarano, H. W. Jensen, P. Hanrahan. *Photon Mapping on Programmable Graphics Hardware.* Graphics Hardware 2003. pp. 41-50, 2003.

[Qureshi, 2000] K. Qureshi, M. Hatanaka. *An introduction to load balancing for parallel raytracing on HDC.* Current Science, Vol. 78, pp. 818-820, No. 7, 10 april 2000.

[Reinhard and al, 2000] E. Reinhard, B. Smits, C. Hansen. *Dynamic acceleration structures for interactive ray tracing*. In Proceedings Eurographics Workshop on Rendering,pages 299–306, Brno, Czech Republic, June 2000.

[Schmittler and al, 2002] J. Schmittler, I. Wald, P. Slusallek. *SaarCOR - A Hardware Architecture for Ray Tracing*. In Proceedings of EUROGRAPHICS Graphics Hardware 2002.

[Wald and al, 2001a] I. Wald, P. Slusallek, C. Benthin, M. Wagner. *Interactive rendering with coherent ray tracing*. In Eurographics 2001.

[Wald and al, 2001b] I. Wald, P. Slusallek, C. Benthin. *Interactive distributed ray tracing of highly complex models*. In Proceedings of the 12th EUROGRPAHICS Workshop on Rendering, June 2001. London.

[Whitted, 1979] J. T. Whitted. *An improved illumination model for shaded display*. ACM Computer Graphics, 13(3):1–14, 1979. (SIGGRAPH Proceedings).

[William and al, 2001] M. William, S. Parker, E. Reinhard, P. Shirley, W. Thompson. *Temporally coherent interactive ray tracing*. Technical Report UUCS-01-005, Computer Graphics Group, University of Utah, 2001.

III

NUMERICAL COMPUTATIONS

EVALUATION OF PARALLEL AGGREGATE CREATION ORDERS: SMOOTHED AGGREGATION ALGEBRAIC MULTIGRID METHOD

Akihiro Fujii
The Center for Continuing Professional Development, Kogakuin University 1-24-2, Nishishinjuku, Shinjuku-ku, Tokyo, Japan; CREST, JST, 4-1-8 Honcho Kawaguchi, Saitama, Japan.
fujii@cpd.kogakuin.ac.jp

Akira Nishida
Department of Computer Science, Graduate School of Information Science and Technology, University of Tokyo 7-3-1, Hongo, Bunkyo-ku, Tokyo, Japan; CREST, JST, 4-1-8 Honcho Kawaguchi, Saitama, Japan.
nishida@is.s.u-tokyo.ac.jp

Yoshio Oyanagi
Department of Computer Science, Graduate School of Information Science and Technology, University of Tokyo 7-3-1, Hongo, Bunkyo-ku, Tokyo, Japan
oyanagi@is.s.u-tokyo.ac.jp

Abstract The Algebraic MultiGrid method (AMG) has been studied intensively as an ideal solver for large scale Poisson problems. The Smoothed Aggregation Algebraic MultiGrid (SA-AMG) method is one of the most efficient of these methods. The aggregation procedure is the most important part of the method and is the main area of interest of several researchers.

Here we investigate aggregate creation orders in the aggregation procedure. Five types of aggregation procedure are tested for isotropic, anisotropic and simple elastic problems. As a result, it is important that aggregates are created around one aggregate in each domain for isotropic problems. For anisotropic problems, aggregates around domain borders influence the convergence much. The best strategy for both anisotropic and isotropic problems in our numerical test is the aggregate creation method which creates aggregates on borders first then creates aggregates around one aggregate in the internal domain.

In our test, the SA-AMG preconditioned Conjugate Gradient (CG) method is compared to the Localized ILU preconditioned CG method. In the experiments, Poisson problems up to 1.6×10^7 DOF are solved on 125PEs.

Keywords: AMG; Poisson Solver; Aggregate Creation

1. Introduction

For large-scale Poisson problems, multi-level methods are known to be efficient solvers. In multi-level methods a smaller problem is constructed from the problem matrix in the setup phase, and this is used to solve the problem in the solution phase. The Smoothed Aggregation Algebraic MultiGrid (SA-AMG) method is one of the most effective multi-level methods. In the SA-AMG method the smaller problem matrices are calculated by creating aggregates of the neighboring unknowns. The convergence efficiency depends on the quality of the aggregates.

The parallel SA-AMG method is often realized by domain decomposition, and parallel aggregation strategies based on domain borders are needed. Thus, parallel aggregation strategies are a very important part of the method, and are discussed in [Ada98, TT00]. These papers propose and compare various aggregation strategies, but they do not consider the order in which the aggregates are created. The order of the aggregate creation also highly influences the convergence. We implement the parallel SA-AMG method which can deal with any aggregation strategy, and investigate which order of aggregate creation is better for both isotropic and anisotropic problems.

Section 2 introduces the SA-AMG method. Section 3 explains the various orders of aggregate creation, and Section 4 summarizes the implementation of the SA-AMG method, followed by numerical experiments and conclusions.

2. SA-AMG Method

The SA-AMG method is one of the MultiGrid methods. First, the MultiGrid method is explained.

It is difficult to solve large-scale problems but the MultiGrid method utilizes more than one grid and solves large-sized problems efficiently. For example, we consider the two grid case of fine and coarse grids. The problem matrices A_1 and A_2 are discretized on the fine grid and the coarse grid respectively. Thus, matrix A_1 is bigger than matrix A_2.

The problem is to solve the equation $A_1x_1 = b_1$, where vector x_1 is the unknown vector. The MultiGrid solution process is as follows.

1 Perform a relaxation, such as the Gauss Seidel method, for the equation $A_1x_1 = b_1$ on the fine grid

2 Compute the residual $r_1 = b_1 - A_1x_1$, and move r_1 to the coarse grid $r_2 = Rr_1$

3 Solve the coarse grid residual equation $A_2x_2 = b_2$ (possibly recursively)

4 Move x_2 to the fine grid and correct the fine grid solution $x_1 \leftarrow x_1 + Px_2$

5 If the convergence criterion isn't satisfied, go to 1

Matrix R represents the restriction operator which moves the vector on the fine grid to the vector on the coarse grid. Matrix P represents the prolongation operator which moves the vector on the coarse grid to the vector on the fine grid. Matrix R is often the transpose of matrix P. Although we consider here a two-level case, Step 3 can be executed recursively and more than two grid cases can be extended to multi-level cases. The MultiGrid method is known to be a very effective method for problems with structured meshes but it is difficult to create coarse grids for unstructured problems. Algebraic MultiGrid (AMG) methods have been studied intensively for such problems. AMG methods create the smaller problem from only the problem matrix without mesh information.

AMG methods have the two phases of setup and solution. The solution phase solves the problem as described above. The setup phase of the AMG method is explained here. The setup phase creates a smaller matrix from the problem matrix. AMG methods often calculate the prolongation matrix P and $R = P^T$ first. Then the smaller problem matrix is calculated from $A_2 = RA_1P$, which is called the Galerkin approximation. If more than one coarse problem matrix is utilized, this process is repeated and the coarser level matrices are calculated in order.

Various AMG methods differ in how the prolongation matrix P is created from the problem matrix A_1. In the SA-AMG method the prolongation matrix P is calculated using the following three procedures.

Filtering the Matrix A_1 Matrix $\tilde{A}_1$ is defined from matrix A_1 by dropping the elements that do not satisfy the following condition.

$$a_{ij}^2 > \theta^2 |a_{ii}||a_{jj}| \tag{1}$$

where θ is a constant value between 1 and 0. $a_{i,j}$ means the element of the i-th row and j-th column of matrix A_1.

Construction of Aggregates We define a graph for the symmetric matrix $\tilde{A}_1$. The vertices and edges in the graph correspond to rows and non-zero elements of matrix $\tilde{A}_1$. The vertex set is divided into disjoint subsets containing neighboring vertices following this procedure.

phase1 : repeat until all unaggregated vertices are adjacent to an aggregate

- pick a root vertex which is not adjacent to any existing aggregate
- define a new aggregate as the root vertex plus all its neighbors

phase2 : sweep unaggregated vertices into existing aggregates or use them to form new aggregates

Figure1 shows the situation. Each subset is called an aggregate, and corresponds to an unknown of the next coarsest level.

Matrix $\tilde{P}$ is an $n \times m$ matrix, where n and m are the numbers of unknowns and aggregates respectively, and is defined as follows:

$$\tilde{P}_{ij} = \begin{cases} 1 & \text{the i-th unknown belongs to the j-th aggregate} \\ 0 & \text{other cases} \end{cases}$$

Relaxation of Aggregates The SA-AMG method defines the prolongation matrix by smoothing the matrix $\tilde{P}$ defined above. Here we assume that the damped Jacobi method is used for smoothing.

$$P = (I - \omega\tilde{D}^{-1}\tilde{A}_1)\tilde{P} \tag{2}$$

where ω and $\tilde{D}$ are the damping coefficient and the diagonal part of $\tilde{A}_1$, respectively. Other methods of smoothing aggregates are described in [VBM01].

Convergence of the SA-AMG method depends significantly on the quality of the multi-level data structure constructed in the setup phase. The construction of aggregates is especially important. For fast convergence, aggregates need to be packed tightly in the domain. As in Figure 2, adjacent aggregates are often created in order. In a parallel computing environment, the problem domain is often decomposed and

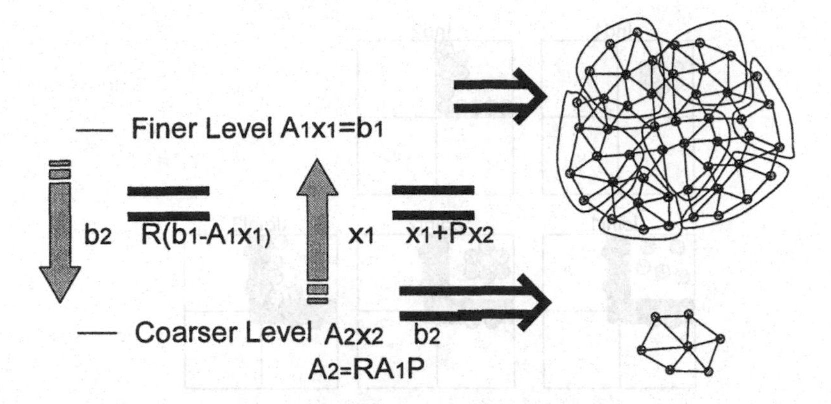

Figure 1. **Conceptual Image of the SA-AMG method: Graph structure is obtained from the matrix**

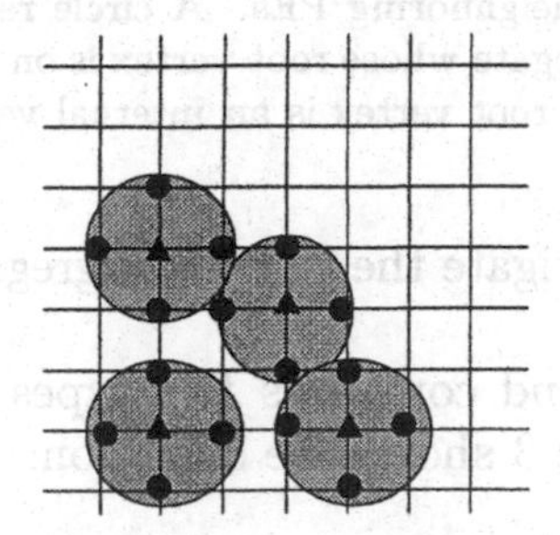

Figure 2. **Aggregation Phase1 for a Five-Point Stencil: The triangle identifies the root vertex and the small circle identifies a vertex neighboring the root vertex**

the aggregation process is executed by each PE. In the next section we discuss strategies for aggregate creation orders for parallel aggregation.

3. Various Orders of Aggregate Creation

Parallel aggregation strategies are discussed in this section. These are classified into two types, independent and joint. Independent aggregation means that the aggregation is carried out independently in each domain. Joint aggregation first makes shared aggregates around borders and then creates aggregates independently in the inner domain.

Adams [Ada98] proposed the Maximal independent set algorithm as an aggregation strategy. It has become one of the joint aggregation strategies. Tuminaro et al. [TT00] proposed another joint aggregation strategy and compared various aggregation strategies. They discussed a method for realizing joint aggregation, but did not investigate aggregate

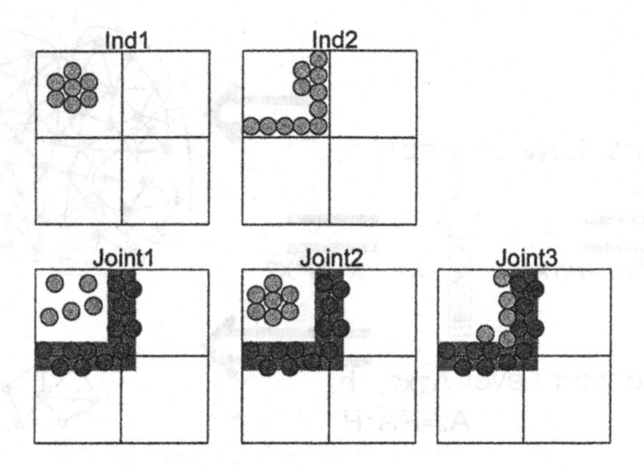

Figure 3. **Two Types of Independent Aggregation: Ind1 and Ind2. Three Types of Joint Aggregation: Joint1, Joint2 and Joint3. The range colored orange represents the vertices on borders with neighboring PEs. A circle represents an aggregate. A green circle represents an aggregate whose root vertex is on a border, and a blue circle represents an aggregate whose root vertex is an internal vertex of the domain.**

creation orders. We investigate the various aggregate creation orders for better convergence.

This paper considers and compares five types of aggregate-creation method as follows. Figure 3 shows the situation.

- Independent Aggregation: Aggregates are created independently from other PEs and aggregates cannot go beyond the borders.
  - Ind1: Aggregates are created around one aggregate in order.
  - Ind2: Aggregates are created from borders in order.
- Joint Aggregation: Root vertices are selected in vertices of borders, greedily at first. Then, root vertices are selected in the interior domain by each PE. The following three strategies differ in the method used to create aggregates in the interior domain.
  - Joint1: Aggregates in the interior domain are created greedily.
  - Joint2: Aggregates in the interior domain are created around one aggregate in order similar to Ind1.
  - Joint3: Aggregates in the interior domain are created around aggregates on borders.

4. Implementation

Our implementation of the SA-AMG method is based on Tuminaro et al. [TT00] and GeoFEM [Geo]. The joint aggregations method is based on Tuminaro et al. [TT00]. Data structure for finite element problems, ICCG solvers and other parts of the implementation are based on GeoFEM.

This section explains the implementation of our solver. The data structure of matrices and the resulting form of the aggregates, which are an important part of the implementation, are discussed.

Firstly, the data structure of the matrices is written. The graph based on the problem matrix is decomposed into sub-domains for PEs. Then, rows and columns of the matrix are reordered by the sub-domains, and the reordered matrix is distributed as a one-dimensional block-row distribution. Each PE has a block of rows from the problem matrix.

Vertices in each PE's domain are also connected with the vertices, called ghost vertices, in the neighboring domain. Each PE's calculation requires the values of the ghost vertices. Thus, communication tables which record the PE number and the vertex number of the ghost vertices are also necessary. Figure 4 shows a simple 2-dimensional finite-element mesh with four PEs. Vertices of the gray part of Figure 4 are required for calculating PE 0's domain, and the vertices 1..25. The ghost vertices are 26..36. $neibPE(:)$, $send(:,p)$ and $recv(:,p)$ have neighboring PE numbers, their own vertices referred to by PE $neibPE(p)$ and the vertex number of the ghost vertices owned by PE $neibPE(p)$.

Secondly, the resulting form of the aggregation is written. Parallel aggregation creates aggregates which are disjoint sets of vertices on the graph based on the problem matrix of each level. It then determines the owner PE for each aggregate. PE k's own aggregates are collected to PE k as vertices on the next coarser level. Thus, the coarser level's domain decomposition follows the determination of the owner PE of the aggregates. The relationship between the owner PE and the aggregates can be any combination, and the domain decomposition of the finer and the coarser levels can be totally changed.

The information for the aggregates is recorded by each PE. There are two types of aggregate recorded by a PE, its own aggregates and external aggregates. A PE's own aggregates are owned by the PE on the next coarsest level. External aggregates are owned by the other PE. Figure 5 shows PE k's 2-dimensional array which contains the vertex numbers of the aggregates. In Figure 5, PE k recorded N_a aggregates. Aggregate numbers $1..N_c$ are the aggregates owned by PE k on the next coarsest level. Aggregate numbers $N_c+1..N_a$ are the aggregates owned

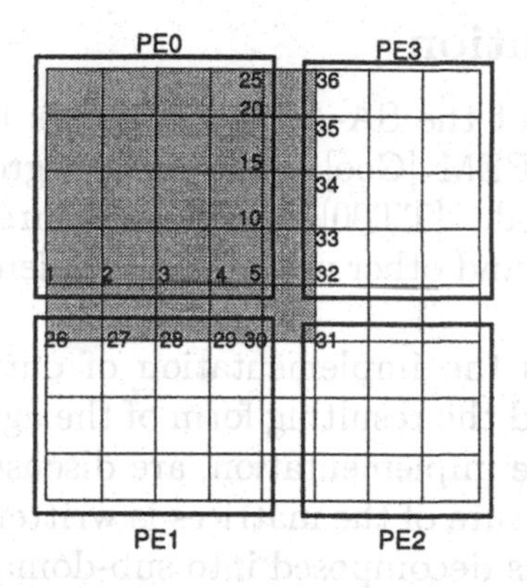

$$
\begin{aligned}
neibPE(:) &= 1,2,3 \\
send(:,1) &= 1,2,3,4,5 \\
send(:,2) &= 5 \\
send(:,3) &= 5,10,15,20,25 \\
recv(:,1) &= 26,27,28,29,30 \\
recv(:,2) &= 31 \\
recv(:,3) &= 32,33,34,35,36
\end{aligned}
$$

Figure 4. Vertices and Communication Tables of PE 0

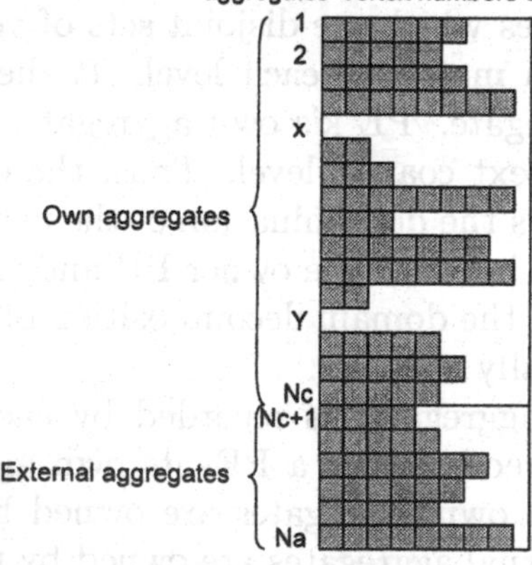

Figure 5. PE k's Aggregate Information Format: Aggregates 1.N_c are owned by PE k on the coarser level. Vertex numbers on PE k's domain are recorded in the array. Aggregates X and Y are owned by PE k on the coarser level, but aggregates X and Y are not on the vertices on PE k's domain.

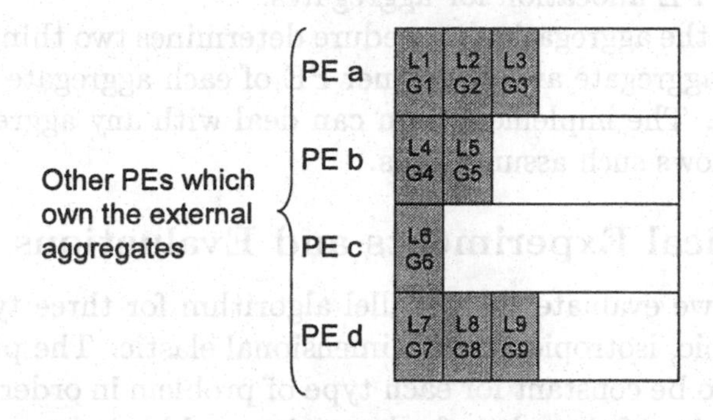

Figure 6. **PE k's Aggregate Table: Aggregate Table records pairs of the global aggregate number (G?) and the local aggregate number (L?) for the external aggregates.**

by other PEs. The array has the aggregates' vertex numbers in PE k's domain. Thus, one aggregate, which contains vertices on the domain of many PEs, is recorded by many PEs. Figure 5 shows that aggregates X and Y have no vertices in PE k's domain and no vertices are recorded. But the two aggregates may have vertices in the other PE's domains and may be recorded as an external aggregate by those PEs. Aggregates are recorded by PE k in the following cases.

- The aggregate's owner PE is PE k.
- The aggregate's vertices contain any vertices of PE k's domain.

Some aggregates are recorded by many PEs, and a global aggregate number is needed to identify them. The local aggregate number is determined and utilized as in Figure 5. The PEs own aggregates are numbered from 1 to the number of owned aggregates (N_c in Figure 5) for each PE. Then, external aggregates follow. In our implementation the global aggregate number is determined by numbering the owned aggregates in the order of PE numbers. For owned aggregates on PE k, the global aggregate number is determined by adding the local aggregate number to the sum of the number of aggregates owned by PEs 0..k-1. External aggregates, however, cannot be identified by a local aggregate number. Therefore, their global aggregate numbers are recorded in an aggregate table as shown in Figure 6. The external aggregate table records the pairs of global and local aggregate numbers for external aggregates.

Ultimately, the output of the aggregation on each PE is recorded as two arrays, sets of vertices (Figure 5) and an aggregate table (Figure 6).

This output interface can deal with any vertex set as an aggregate and can cope with any PE allocation for aggregates.

We assume that the aggregation procedure determines two things: the vertex set of each aggregate and the owner PE of each aggregate on the next coarsest level. The implementation can deal with any aggregation strategy which follows such assumptions.

5. Numerical Experiments and Evaluations

In this section, we evaluate the parallel algorithm for three types of problem: anisotropic, isotropic, and 3-dimensional elastic. The problem size per PE is set to be constant for each type of problem in order to understand the behavior of the solver for large-size problems on massively parallel systems.

The SA-AMG mathod and Localized ILU preconditioned CG method [NO99, NNT97, Nak03], which is refered to as ICCG, are compared in the experiments. In addition, five aggregation strategies for the SA-AMG method are tested. These are introduced in section 3.

The SA-AMG method utilizes a V-cycle for the solution phase. One iteration of the Chaotic Symmetric Gauss Seidel(Chaotic SGS) method is carried out as pre- and post-smoothing at each level. At the coarsest level, twenty iterations of the Chaotic SGS method or parallel direct solution method are performed. We utilize PSPASES [GGJ+97] as the parallel direct solver.

Numerical Experiment Environment Numerical experiments are carried out on a cluster with 128 nodes of Sun Blade1000 workstations which have dual CPUs of UltraSPARC III 750MHz and 1GB memory. These nodes are connected by Myrinet2000. For parallelization, MPICH-GM [MPI, Myr] is used. The code is written in Fortran90.

5.1 Anisotropic Problems

Equation 3 with a rectangular parallelepiped domain discretized by a finite element method is solved for anisotropic and isotropic problems. The problem domain has six surfaces perpendicular to the x, y and z axes. In Equation 3 Xmin, Ymin, Zmax and Zmin represent surfaces with minimum X value, minimum Y value, maximum Z value and minimum Z value respectively. In the anisotropic case, ϵ in equation 3 is set to 0.0001. This anisotropic problem is very difficult to solve, from the view point of condition number.

Table 1. **Problem Size and Number of PEs for Anisotropic Problems: DOF represents Degree of Freedom of the problem.**

Anisotropic Problems		
DOF	# of PEs	Problem Domain
8192k	128	160 × 160 × 320
4096k	64	160 × 160 × 160
2048k	32	80 × 160 × 160
1024k	16	80 × 80 × 160
512k	8	80 × 80 × 80
256k	4	40 × 80 × 80
128k	2	40 × 40 × 80

$$\partial/\partial x(\partial p/\partial x) + \partial/\partial y(\partial p/\partial y) + \partial/\partial z(\epsilon\, \partial p/\partial z) = 0 \qquad (3)$$

$$\begin{cases} Xmin : \partial p/\partial x = 10.0 \\ Ymin : \partial p/\partial y = 5.0 \\ Zmax : \partial p/\partial z = 1.0 \\ Zmin : p = 0.0 \end{cases}$$

The problem domain is decomposed into sub-domains with $40\times 40\times 40$ vertices, which are allocated to each PE. Table 1 shows the size of the problems and the number of PEs. The convergence criterion is that the 2-norm of the relative residual is less than 10^{-7}.

For anisotropic problems, the coarsest level problem is still difficult to solve. We tested some aggregation strategies with a parallel direct solver at the coarsest level. There are SA-AMG methods with five levels and four levels. SA-AMG methods with five levels utilize twenty iterations of the Chaotic SGS method at the coarsest level, and SA-AMG methods with four levels utilize the parallel direct solver at the coarsest level. In Table 2, the problem size at the coarsest level is described for each aggregation strategy. The problem size is biggest in the 128PE case and that case is written.

ICCG and SA-AMG Methods Tables 3 and 4 show the total times for the ICCG and SA-AMG methods with Ind1 of independent aggregation for anisotropic problems. The ICCG method doesn't reach the convergence criterion for the biggest problem and indicates the difficulty of the problem. On the other hand, Table 4 shows that the SA-AMG method works well for anisotropic problems in comparison with

Coarsest Level Size for 128PE case		
Aggregation Strategy	Size	# of Levels
Ind1	7864	5
Ind2	6375	5
Ind2 with Parallel Direct Solver	22601	4
Joint1	2602	5
Joint2	3328	5
Joint3	3136	5
Joint1 with Parallel Direct Solver	20524	4

Table 2. Coarsest Problem Size for Anisotropic Problems

the ICCG method. The following paragraphs compare and evaluate aggregation strategies for the SA-AMG method for anisotropic problems.

Adaptation of Independent Aggregation for Anisotropy Tables 4, 5, and 6 have the iteration number, time of each phase, and the total time for the SA-AMG method with three types of independent aggregation strategy. Table 4 shows the normal independent aggregation of Ind1. This aggregation makes irregular aggregates around borders, and the iteration number for convergence increases. Table 5 shows the result of aggregation from borders, which is explained as Ind2 of independent aggregation in section 3. This improvement does well in convergence. In addition, the number of levels is reduced with the parallel direct solver at the coarsest level. The result is shown in Table 6. By using these improvements, independent aggregation can be adapted to anisotropic problems.

Joint Aggregation for Anisotropy Tables 7, 8, 9 and 10 show the results of various aggregation strategies for joint aggregation, Joint1, Joint2, Joint3, and Joint1 with fewer levels and a parallel direct solver. Joint1, Joint2 and Joint3 differ in their aggregate creation method for the inner part of each PE's domain. Joint1 creates aggregates in the order of the vertex numbers, and Joint2 creates aggregates around one aggregate in order. Joint3 creates aggregates from borders. These are further explained in section 3. Their performances for anisotropic problems differ little. It also shows that improvements in the reduction of level number and the use of a parallel direct solver, work well. In comparison with independent aggregation, joint aggregations works effectively for anisotropic problems.

ICCG method		
# of PEs	# of Iteration	Total Time
128	>10000	> 2106
64	5527	1169.7
32	5426	1100.4
16	5232	1030.2
8	2750	530.1
4	2677	506.0
2	2219	412.1

Table 3. ICCG Method for Anisotropic Problems. Time is written in seconds.

Aggregation Strategy For Anisotropic Problems For this numerical experiment, joint aggregation with a parallel direct solver is the most efficient. Ind2 of independent aggregation with a parallel direct solver, however, works nearly as well as the previous method. Except for methods using a parallel direct solver, any joint aggregation works well for anisotropic problems.

Ind1 of Independent Aggregation				
# of PEs	# of Iteration	Setup Time	Solution Time	Total Time
128	515	5.7	443.0	448.8
64	732	5.6	622.2	627.9
32	506	5.0	399.5	404.5
16	73	4.3	51.8	56.2
8	47	4.2	32.1	36.4
4	60	3.7	39.4	43.1
2	27	3.3	16.9	20.4

Table 4. SA-AMG Method for Anisotropic Problems: Ind1 of independent aggregation creates aggregates around one aggregate in order. The number of levels is five. Time is written in seconds.

5.2 Isotropic Problems

For isotropic problems, ϵ in equation 3 is set to 1. Problem domain, discretization method and boundary conditions are the same as the anisotropic case. The problem domain is decomposed into sub-domains with $50 \times 50 \times 50$ vertices, which are allocated to each PE. Table 11 shows the sizes of problems and the numbers of PEs. The convergence criterion is that the 2-norm of the relative residual is less than 10^{-12}.

For isotropic problems, all SA-AMG methods have four levels and utilize twenty iterations of the Chaotic SGS method at the coarsest level.

Ind2 of Independent Aggregation				
# of PEs	# of Iteration	Setup Time	Solution Time	Total Time
128	245	5.0	205.6	210.7
64	147	5.0	122.6	127.6
32	65	4.6	50.0	54.7
16	62	4.1	42.9	47.1
8	43	3.9	28.9	32.9
4	44	3.5	28.4	31.9
2	29	3.3	18.2	21.6

Table 5. **SA-AMG Method for Anisotropic Problems: Ind2 of independent aggregation creates aggregates from borders. The number of levels is five. Time is written in seconds.**

Ind2 of Independent Aggregation with fewer levels and a parallel direct solver				
# of PEs	# of Iteration	Setup Time	Solution Time	Total Time
128	82	6.9	62.1	69.1
64	34	5.8	24.9	30.8
32	33	4.9	23.1	28.1
16	26	4.2	17.4	21.6
8	26	4.0	17.0	21.1
4	22	3.5	14.1	17.7
2	21	3.3	13.3	16.6

Table 6. **SA-AMG Method for Anisotropic Problems: Ind2 of independent aggregation with fewer levels utilizing a parallel direct solver at the coarsest level. The number of levels is four. Time is written in seconds.**

Joint1 of Joint Aggregation				
# of PEs	# of Iteration	Setup Time	Solution Time	Total Time
128	99	4.9	78.1	83.1
64	59	4.6	46.7	51.4
32	48	4.2	36.1	40.4
16	47	4.0	32.2	36.2
8	32	3.8	21.4	25.2
4	20	3.4	12.8	16.3
2	26	3.2	16.2	19.5

Table 7. **SA-AMG Method for Anisotropic Problems: Joint1 of joint aggregation is explained in section 3. Aggregates in the internal domain are created greedily. The number of levels is five. Time is written in seconds.**

Joint2 of Joint Aggregation				
# of PEs	# of Iteration	Setup Time	Solution Time	Total Time
128	109	5.1	89.4	94.6
64	64	5.0	52.5	57.5
32	51	4.5	37.9	42.5
16	48	4.2	33.2	37.5
8	32	4.0	22.4	26.4
4	26	3.6	16.8	20.4
2	32	3.3	20.1	23.5

Table 8. SA-AMG Method for Anisotropic Problems: Joint2 of joint aggregation is explained in section 3. Aggregates in the internal domain are created around one aggregate in order. Time is written in seconds.

Joint3 of Joint Aggregation				
# of PEs	# of Iteration	Setup Time	Solution Time	Total Time
128	102	5.1	83.9	89.1
64	62	4.9	50.6	55.6
32	51	4.5	38.0	42.6
16	47	4.2	32.2	36.6
8	31	4.0	20.8	24.9
4	26	3.6	16.6	20.3
2	32	3.4	20.2	23.6

Table 9. SA-AMG Method for Anisotropic Problems: Joint3 of joint aggregation is explained in section 3. Aggregates in the internal domain are created around aggregates on borders in order. Time is written in seconds.

Joint1 of Joint Aggregation with fewer levels and parallel direct solver				
# of PEs	# of Iteration	Setup Time	Solution Time	Total Time
128	76	6.9	57.1	64.1
64	40	5.2	29.0	34.2
32	18	4.5	12.7	17.3
16	15	4.0	10.1	14.2
8	15	3.8	9.9	13.8
4	15	3.5	9.6	13.1
2	18	3.2	11.3	14.5

Table 10. SA-AMG Method for Anisotropic Problems: Joint1 of joint aggregation is improved with fewer levels utilizing a parallel direct solver at the coarsest level. The number of levels is four. Time is written in seconds.

Table 11. Problem Size and Number of PEs for Isotropic Problems

Isotropic Problems		
DOF	# of PEs	Problem Domain
15625k	125	250 × 250 × 250
12500k	100	200 × 250 × 250
10000k	80	200 × 200 × 250
8000k	64	200 × 200 × 200
6000k	48	150 × 200 × 200
4500k	36	150 × 150 × 200
3375k	27	150 × 150 × 150
2250k	18	100 × 150 × 150
1500k	12	100 × 100 × 150
1000k	8	100 × 100 × 100
500k	4	50 × 100 × 100
250k	2	50 × 50 × 100

Coarsest Level Size for the 125PE case		
Aggregation Strategy	Size	# of Levels
Ind1	2457	4
Ind2	1562	4
Joint1	986	4
Joint2	1514	4
Joint3	1048	4

Table 12. Coarsest Problem Size for Isotropic Problems

When the SA-AMG method utilizes Ind1 of independent aggregation in 125PE case, one PE has 125000, 9319, 526, 20 unknowns for each level.

In Table 12, the problem size at the coarsest level is described for each aggregation strategy. The problem size is the biggest for the 125PE case and that case is written.

ICCG and SA-AMG Methods The results of the ICCG method for isotropic problems are shown in Table 13. The ICCG method's total time increases with problem size. On the other hand, Tables 14, 15, 16 17 and 18 show the result of the SA-AMG methods with various aggregations. In comparison with the ICCG method, Tables of the SA-AMG methods show that they need an almost constant time until convergence for any problem size on the grounds that problem size per PE is constant in this experiment. The SA-AMG methods work well especially for large-sized problems.

ICCG method		
# of PEs	# of Iteration	Total Time
125	602	267.9
100	590	260.0
80	537	235.2
64	481	210.5
48	463	202.4
36	417	180.8
27	364	157.8
18	340	145.7
12	296	124.9
8	241	100.9
4	207	86.6
2	165	67.3

Table 13. ICCG Method for isotropic Problems. Time is written in seconds.

Ind1 of Independent Aggregation				
# of PEs	# of Iteration	Setup Time	Solution Time	Total Time
125	23	11.1	39.1	50.2
100	23	10.9	38.4	49.4
80	22	10.8	36.8	47.8
64	22	10.8	36.6	47.5
48	22	10.7	36.6	47.4
36	22	10.7	36.2	46.9
27	21	10.4	34.4	44.8
18	21	10.2	33.5	43.8
12	21	9.8	32.5	42.4
8	21	9.4	31.6	41.1
4	21	8.8	30.7	39.6
2	21	8.4	29.6	38.0

Table 14. SA-AMG Method for Isotropic Problems: Ind1 of independent creates aggregates around one aggregate in order. Time is written in seconds.

Ind2 of Independent Aggregation				
# of PEs	# of Iteration	Setup Time	Solution Time	Total Time
125	37	10.2	61.3	71.5
100	35	10.0	56.6	66.7
80	35	10.0	56.2	66.4
64	35	10.0	55.8	65.8
48	35	10.0	55.5	65.6
36	34	9.9	53.2	63.1
27	34	9.9	52.6	62.5
18	34	9.6	51.3	61.0
12	33	9.2	48.5	57.8
8	31	8.9	44.9	53.9
4	28	8.5	39.9	48.4
2	25	8.1	34.9	43.1

Table 15. SA-AMG Method for Isotropic Problems: Ind2 of independent aggregation creates aggregates from borders. Time is written in seconds.

Joint1 of Joint Aggregation				
# of PEs	# of Iteration	Setup Time	Solution Time	Total Time
125	62	10.6	110.2	120.9
100	62	10.4	106.9	117.4
80	61	10.5	104.4	114.9
64	60	10.3	101.2	111.6
48	59	11.0	98.0	109.1
36	59	10.3	97.9	107.3
27	57	10.0	91.8	101.8
18	55	9.5	86.2	95.9
12	55	9.2	84.4	93.7
8	51	9.0	77.1	86.1
4	51	8.8	76.9	85.7
2	48	7.8	69.1	77.0

Table 16. SA-AMG Method for Isotropic Problems: Joint1 of joint aggregation creates aggregates in the order of the vertex number in the inner domain. Time is written in seconds.

Joint2 of Joint Aggregation				
# of PEs	# of Iteration	Setup Time	Solution Time	Total Time
125	28	10.9	48.4	59.4
100	28	10.7	47.8	58.6
80	28	10.6	47.2	57.9
64	27	10.5	45.1	55.7
48	27	10.5	44.7	55.2
36	27	10.2	43.9	54.2
27	26	10.5	41.6	52.2
18	25	10.9	39.1	49.3
12	25	9.7	38.2	48.0
8	23	9.4	34.7	44.1
4	22	8.9	32.0	40.9
2	19	8.4	27.0	35.5

Table 17. **SA-AMG Method for Isotropic Problems: Joint2 of joint aggregation creates aggregates around one aggregate in order in the inner domain. Time is written in seconds.**

Joint3 of Joint Aggregation				
# of PEs	# of Iteration	Setup Time	Solution Time	Total Time
125	55	11.6	95.8	107.5
100	55	11.4	93.6	105.1
80	55	11.3	92.9	104.2
64	54	11.2	89.4	100.7
48	54	11.2	89.1	100.3
36	53	10.9	85.2	96.2
27	50	10.7	78.6	89.4
18	51	10.6	79.2	89.9
12	49	10.2	74.4	84.7
8	44	9.8	65.9	75.8
4	50	9.8	74.5	84.3
2	37	8.9	52.4	61.4

Table 18. **SA-AMG Method for Isotropic Problems: Joint3 of joint aggregation creates aggregates from borders in the inner domain. Time is written in seconds.**

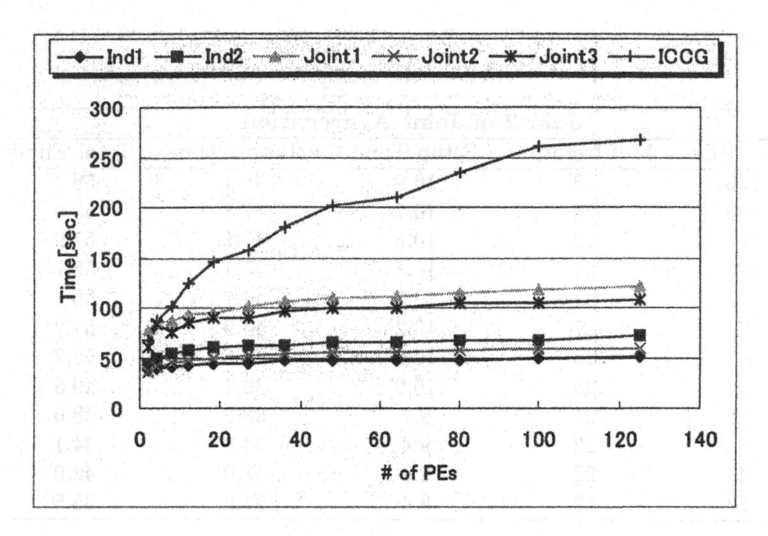

Figure 7. ICCG and SA-AMG Methods for Isotropic Problems: Ind1, Ind2, Joint1, Joint2 and Joint3 represent the aggregation methods of the SA-AMG method explained in section 3.

The SA-AMG method with Various Aggregations Unlike the anisotropic cases, Ind2 of independent aggregation from borders is inferior to Ind1 of normal independent aggregation for isotropic problems, according to Tables 14 and 15. In comparison with other aggregation strategies, Ind1 is the best strategy for these problems. The iteration numbers of Ind1 are almost constant. These numbers are between 21 and 23 for problems whose sizes are from 2.5×10^5 DOF and 1.56×10^7 DOF.

Next we consider joint aggregation strategies. Joint aggregations create aggregates on borders, then create the inner part of the domain. Joint1, Joint2 and Joint3 differ in the creation of aggregates in the inner part of the domain. Joint1 creates aggregates in the order of aggregate number, which corresponds to a greedy algorithm. Joint2 creates aggregates around one aggregate in order. Joint3 creates aggregates from borders. Tables 16, 17, 18 show the results of joint aggregation. Unlike the anisotropic problems, joint aggregations differ considerably in performance. Joint2 is the best among the three methods.

Figure 7 shows the total time for the ICCG and SA-AMG methods for each problem size. It shows the Ind1 of the normal independent aggregation strategy requires the shortest total time in our experiments. Ind1 and Joint2 of the aggregation strategies work relatively well.

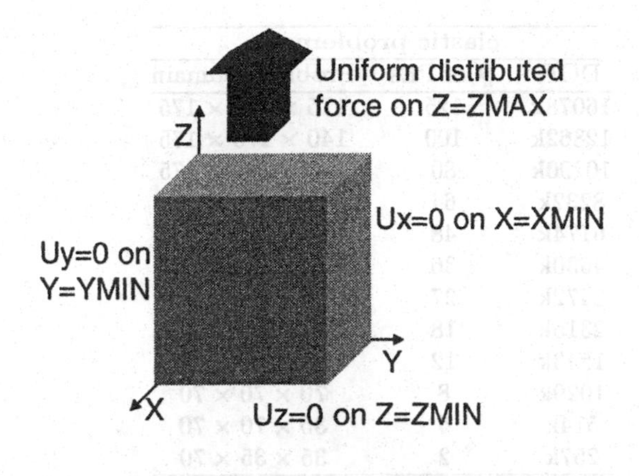

Figure 8. Cube

5.3 3-Dimensional Problems in Elasticity

The problem of the pulled cube is one of the test problems for GeoFEM. The problem is to compute the displacement U of a cube with surfaces perpendicular to the x, y and z axes. U_x represents the displacement U in the direction of the x-axis. The surface of Z=ZMAX is pulled by a uniform distributed force in the Z direction. The other boundary conditions are $U_x = 0$ on the surface of X=XMIN, $U_y = 0$ on the surface of Y=YMIN, and $U_z = 0$ on the surface of Z=ZMIN. Figure 8 shows the situation. The problem domain is decomposed into sub-domains with $35 \times 35 \times 35$ vertices, which are allocated to each PE. A vertex of finite element mesh has three DOF, corresponding to displacements in the x, y and z directions. Thus each PE deals with $35 \times 35 \times 35 \times 3$ DOF. The ICCG and SA-AMG methods are 3×3 blocked for this problem.

There is little difference in performance among parallel aggregation strategies for this problem. It seems that the problem size for each PE is too small for parallel aggregation strategies to make a difference. In this subsection, comparison between the SA-AMG and ICCG methods, and the behavior of the SA-AMG method for large problems in elasticity, are investigated. For elastic problems, all SA-AMG methods have four levels and utilize twenty iterations of the Chaotic SGS method at the coarsest level. When the SA-AMG method utilizes normal independent aggregation for the 125PE case, one PE has 42875×3, 1585×3, 80×3, and 5×3 unknowns for each level.

ICCG and SA-AMG Methods This paragraph compares the ICCG and SA-AMG methods for problems in elasticity. Figure 9 shows the

elastic problems		
DOF	# of PEs	Problem Domain
16078k	125	$175 \times 175 \times 175$
12862k	100	$140 \times 175 \times 175$
10290k	80	$140 \times 140 \times 175$
8232k	64	$140 \times 140 \times 140$
6174k	48	$105 \times 140 \times 140$
4630k	36	$105 \times 105 \times 140$
3472k	27	$105 \times 105 \times 105$
2315k	18	$70 \times 105 \times 105$
1543k	12	$70 \times 70 \times 105$
1029k	8	$70 \times 70 \times 70$
514k	4	$35 \times 70 \times 70$
257k	2	$35 \times 35 \times 70$

Table 19. **Problem Size and Number of PEs for Elastic Problems**

total time for the ICCG and SA-AMG methods for each problem size. ICCG method's total time increases along with the problem size. On the other hand, the SA-AMG method's total time increases little as problem size increases. This means that the number of iterations for the SA-AMG method until convergencen is almost constant. For example, the number of iterations for the SA-AMG method with Ind2 of independent aggregation is 48 for 2 PEs and 52 for 125 PEs. For this problem, the SA-AMG methods work better than ICCG method for all problem sizes. The differences in performance of aggregation strategies are small, partly because the vertex size $35 \times 35 \times 35$ of the problem allocated to each PE is small in comparison with isotropic and anisotropic problems.

6. Summary and Conclusions

This paper compares various aggregation strategies seeking robust strategies for both isotropic and anisotropic problems. We implement the SA-AMG method which can deal with any aggregation strategy and solve anisotropic, isotropic and elastic problems. There are two contributions. First is that independent aggregation can be adapted to anisotropic problems by creating aggregates from borders and utilizing a parallel direct solver at the coarsest level. Independent aggregation is known to be a bad aggregation strategy, as is shown in [TT00]. Second is that the robust aggregation strategy is a strategy that creates aggregates around one aggregate in order after shared aggregates are created on borders. It is explained as Joint2 in section 3. The results for each type of problem are given in subsequent paragraphs.

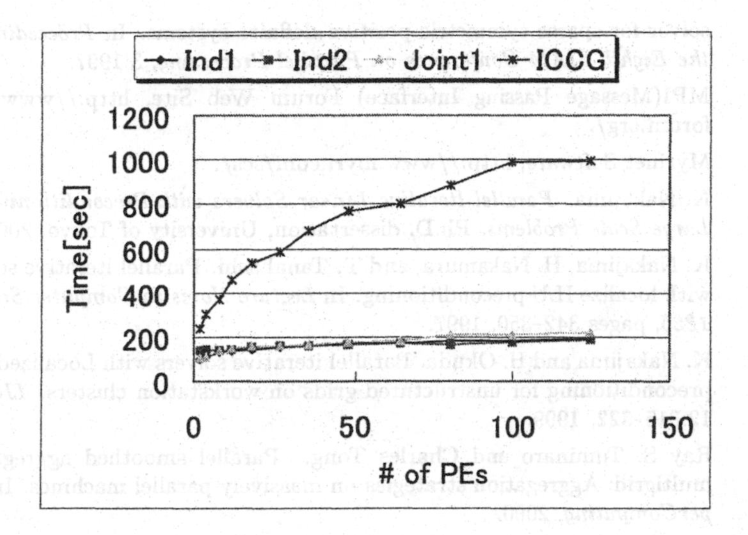

Figure 9. ICCG and SA-AMG Methods for Elastic Problems: Ind1, Ind2, and Joint1 represent aggregation strategies for the SA-AMG method.

For anisotropic problems, joint aggregations work more efficiently than independent aggregations, but Ind2 of independent aggregation with a parallel direct solver works as well as joint aggregations with a parallel direct solver. Methods such as Ind2 or joint aggregations create aggregates around borders first. That aggregate creation order creates no distorted aggregates around borders, which seems to be important for anisotropic problems.

For isotropic problems, Ind1 of independent aggregation is the best aggregation strategy in the total time. Joint aggregations differ in their performance much more than the anisotropic case. Joint2 of joint aggregation works the best of the joint aggregations. Ind1 and Joint2 create aggregates around one aggregate in order in the internal domain. This aggregate creation order in the internal domain seems to be important for isotropic problems. For elastic problems, there is little difference in performance. There seems to be too small a domain allocated for each PE.

References

[Ada98] M. F. Adams. A parallel maximal independent set algorithm. In *Proceedings 5th Copper mountain conference on iterative methods*, 1998.

[Geo] GeoFEM, http://www.geofem.tokyo.rist.or.jp/.

[GGJ+97] Anshul Gupta, Fred Gustavson, Mahesh Joshi, George Karypis, and Vipin Kumar. Design and implementation of a scalable parallel direct solver for sparse symmetric positive definite systems. In *Proceedings of the Eighth SIAM Conference on Parallel Processing*, 3 1997.

[MPI] MPI(Message Passing Interface) Forum Web Site, http://www.mpi-forum.org/.

[Myr] Myrinet Software, http://www.myri.com/scs/.

[Nak03] K. Nakajima. *Parallel Iterative Linear Solvers with Preconditioning for Large Scale Problems.* Ph.D. dissertation, University of Tokyo, 2003.

[NNT97] K. Nakajima, H. Nakamura, and T. Tanahashi. Parallel iterative solvers with localize ILU preconditioning. In *Lecture Notes in Computer Science 1225*, pages 342–350, 1997.

[NO99] K. Nakajima and H. Okuda. Parallel iterative solvers with Localized ILU preconditioning for unstructured grids on workstation clusters. *IJCFD*, 12:315–322, 1999.

[TT00] Ray S. Tuminaro and Charles Tong. Parallel smoothed aggregation multigrid: Aggregation strategies on massively parallel machines. In *SuperComputing*, 2000.

[VBM01] Petr Vanek, Marian Brezina, and Jan Mandel. Convergence of algebraic multigrid based on smoothed aggregation. *Numerische Mathematic*, 88(3):559–579, 2001.

PINPOINTING THE REAL ZEROS OF ANALYTIC FUNCTIONS

Soufiane Noureddine
Dept. of Mathematics and Computer Science, University of Lethbridge
4401 University Drive, Lethbridge, T1K 3M4, AB, Canada
soufiane.noureddine@uleth.ca

Abdelaziz Fellah
Dept. of Computer Science, University of Sharjah
P.O. Box: 27272, Sharjah, United Arab Emirates
fellah@sharjah.ac.ae

Abstract We present a new algorithm for extracting the real zeros of an analytic function. The algorithm makes use of known results of analytic function theory. The distinguishing feature of the algorithm is its ability to handle only zeros the analyst is interested in and to work with sufficient accuracy in cases where the number of zeros is extremely large. In the course of our discussion, we introduce two novel concepts: sweeping functions and size of polynomials.

Keywords: Zeros of analytic functions, Principle of argument, Bisection methods, Quadrature.

1. Introduction and Motivation

Perhaps there is no topic in applied mathematics that has occupied scientists more than solving equations. The task of solving equations concerned not only (pure) mathematicians, but also computer scientists, physicists, engineers, and scientists in many other disciplines. In computer science, for example, the topic is strongly emphasized in numerical methods of computing, and makes itself well visible in different application domains such as robotics, computer graphics, and computer vision, only to mention some.

One might ask now, why introduce a new method for finding zeros? In fact, the method presented here addresses a usually unmentioned and/or underestimated problem: *What are the values of the zeros we are interested in and how many are they*? It turns out that both questions are not easily answered. As a matter of fact, all known methods for zero finding will contribute only to a certain extent to the solution of both problems. The trouble in working with well-known methods arises partly because (a) They cannot solve your problem. (b) They could solve your problem in some instances only. (c) They can solve your problem but you are misusing these methods. Following this classification we can easily notice that known methods are either of type (a) or (b) regarding the problem of finding real zeros of an analytic function. Methods that easily work if only real zeros are needed is therefore obvious.

Restrictions on the kind of zeros a user may be interested in is natural in real applications. For example, Henrici (Henrici, 1974) mentioned the stability of polynomials in control theory, which translates in zeros having negative real parts. The equation for modeling heat propagation (Greenleaf, 1972) is another example for an application where one is only interested in the real zeros contained in some domain.

We will restrict our discussion to real zeros of *analytic* functions. The rationale behind this assumption is that analytic functions are very common in practice and fortunately very well understood. We will be able, therefore, to make extensive use of the complex analysis results related to analytic functions. The problem statement of this paper can be stated as follows:

Given an analytic function $f(z)$, where z is a complex variable, under which conditions is it possible to locate the real zeros of $f(z)$ in a given real interval $[a, b]$? If said conditions are satisfied, then find an efficient and reliable algorithm that locates the real zeros in $[a, b]$ and/or their number.

An important aspect in this connection is that we are interested in *real zeros* (contained in some domain) only. There are algorithms that solve the same problem without the restriction to real zeros (Delves and Lyness, 1967(a); Lehmer, 1962). In fact, the intention of both algorithms is to determine good approximations for all zeros of an analytic function in a systematic way (but not simultaneously). In many applications, the function under study would be modeling some real world behavior and in many instances the pure complex zeros of $f(z)$ will not bear any meaning with regard to the modeled behavior. Thus, it is worthwhile to think about a general method for this type of applications.

The second assumption is liberating. We will omit the usual assumption tacitly made in this kind of work that the number of zeros N is tractable, that is, N is not too large. This will restrict our solution space even further but the authors are convinced that this step is needed in many cases (this will become clearer as we proceed in the paper). Other works like that of (Delves and Lyness, 1967(a)) and those based thereupon actually address the tractability of the number of zeros in a given domain, too. But the main solution approach there is bisection. The domain under examination is shortened step by step until it only includes a small number of zeros. Then, some direct method is used to locate the zeros, for example, Delves and Lyness (Delves and Lyness,

1967(a)) used polynomial root-finding. Bearing in mind that these methods attempt to locate *a*ll the zeros of the given function $f(z)$, it is obvious that the bisection algorithm, though extremely powerful for finding few zeros, is of no help if the number of zeros N is too high, *i.e.*, $N = 2^k$ for some k.

The price for the last assumption is that in our case it is not reasonable anymore to ask for the locations of all zeros. We will have rather to content ourselves with the determination of few zeros in question, or even coarser, with the mere information that the given interval contains at least one real zero. Obviously, if we can assure by some means that the number of zeros in the respective interval is not too large then precisely determining their locations can be easily achieved using a bisection scheme. Thus, we will not exclude the simpler case where N is relatively small in our discussion.

The paper is organized as follows. In the next section, we review some of the well-known zero finding methods. In Section 3, we sketch the general principles of our solution approach in a generic way. The formal algorithm of pinpointing the real zeros of analytic function is presented in Section 4. Section 5 reports on some experiments obtained in a prototypical implementation of the algorithm. Finally, in Section 6 we briefly elaborate on further investigations and conclude the paper with some critical remarks.

2. Related Work

Without claiming to be exhaustive, let us very briefly review some of the well-known zero finding methods and assess them with respect to our problem. Roughly, these methods can be classified as follows:

Rule of Signs for Polynomials

The Descartes rule of signs (see e.g., (Henrici, 1974)) is one of the oldest methods to quickly get a first rough upper bound of the number of real zeros of real polynomials (and easily generalizes to other entire real-valued functions with real coefficients). Sometimes the rule is used to get lower bounds for non-real zeros, too (Drucker, 1979). This rule simply says that the number of real zeros (counting multiplicities) in a real interval $[a, b]$ is either the number of coefficients sign changes, say C, of a certain polynomial (gained from the original one) or C minus an even number. Combined with a bisection scheme, for example, the method can be extended to a technique for the determination of real zeros of polynomials (e.g., using Vincent's algorithm (Vincent)). The main problem of this method, however, is that it is based on upper bounds of the number of zeros. Thus, the method is only conditionally applicable (in the sense of (Henrici, 1974)). A nice feature of this rule, however, is that the complexity of the determination of this upper bound *does not depend on the degree of the polynomial*. This feature however has no major practical relevance, since a bisection scheme potentially leads to a high number of coefficients anyway. Better is to resort to Sturm sequences (Henrici, 1974) which takes $O(n^2)$ steps where n the degree of the input polynomial. Although exact, Sturm sequences are inefficient if n is too large.

Iterative Methods

One clearly thinks here of Fixed-Point methods with the well-known representative, namely, Newton's method (Henrici, 1974; Douglas and Burden, 2003). The latter has the nice feature of being quadratically convergent in contrast to the general fixed-point method, whose (potential) convergence is only linear. Newton's method converges locally only. That is, the first guess of the zero should be good enough. Rather than heuristically guessing the value of the zero, advanced methods use Newton's method only in the late stages of the zero finding process. Newton's method is not always convergent and converges to real zeros only, if the initial guess is real. The need for the derivative makes Newton's method sometimes inconvenient. Related methods are the Laguerre's method (Henrici, 1974), which may cubically converge to complex zeros even with a real starting value. If all zeros are real (e.g., eigenvalues of a symmetric matrix), the method is even globally convergent (Foster, 1981). However, Laguerre's method and more general methods like Schroeder's iteration methods (Henrici, 1974), which have arbitrarily large convergence rates, need more function evaluations per iteration, a fact that makes them less useful in practice. There are also iterative methods that determine *all* zeros of an analytic function simultaneously. Our method is somehow completely different, in the sense that we only select some of the zeros to be determined. Again, the main idea is to have an initial guess that is refined iteratively. If all zeros are known to be simple (Petcovic et. al., 1995) gives an iterative method that uses numerical integration for the determination of all zeros. Other similar methods have been presented in (Petcovic and Marjanovic, 1991; Atanassova, 1994; Olver, 1952).

Bisection Methods

The main idea of these methods is already present in simple algorithms like Regula Falsi and Secant methods (see any numerical analysis textbook, e.g., (Douglas and Burden, 2003)). They always converge to the bracketed zeros (if any). More general algorithms work equally well in the field of complex numbers (and other fields as well). Henrici (Henrici, 1974) provided a simple algorithm for determining the winding number of a closed Jordan curve in the complex plane. His algorithm, though only conditionally reliable, is the basis of many bisection methods for zero determination. Lehmer (Ying and Katz, 1988) used a direct numerical quadrature for the determination of the number of zeros of an analytic function. He combined his basic algorithm with a bisection method that is based on circular contours. Many other known algorithms are based on the Lehmer's approach. The best representative is that of (Delves and Lyness, 1967(a); Delves and Lyness, 1967(b)). They handled the zero finding problem in a rigorous manner, used a scheme akin to Lehmer's method with an emphasis on entailed numerical errors, treated circles and squares as contours, and also provided a derivative-free method based on the ideas of Henrici. Li (Li, 1983) gave later another variant of Delves and Lyness method using homotopy continuation methods. Herlocker and Ely (Herlocker and Ely, 1995) exploited automatic differentiation and interval arithmetic to bound the error of numerical quadrature for the determining the zeros. An excellent treatment for the zeros of analytic functions which is based on numerical quadrature can be found in (Kravanja, Barel and Van, 2000).

For the above mentioned reason, our algorithm shall be based on a bisection scheme that exploits numerical quadrature. All known algorithms go into trouble if N (the number of zeros counting multiplicities) is too large. Our algorithm tries to mitigate this problem found in other algorithms. Also, the algorithm in its current form is able to extract the real zeros of an analytic function with relatively high accuracy. This feature is not found in other known algorithms. Moreover, under some circumstances the presented concepts generalize easily to the extraction of zeros contained in any bounded domain of the complex plane. It will be explained in the course of our treatment that the solution needs the (numerical) evaluation of some singular integrals. The reader should, however, keep in mind that we cannot make use of the available arsenal of methods that work well for this type of integrals for the simple reason that those methods assume the explicit knowledge of the singularities in question. Unfortunately, the singularities in our problem are the zeros to be evaluated, so they cannot be assumed to be known in advance.

3. Assumptions and Approach

Our approach will be based on the fundamental concept of complex analysis, that is, the so-called *Principle of Argument.* See for example, (Kravanja, Barel and Van, 2000 ; Delves and Lyness, 1967(a); Lehmer, 1962). The principle is stated as follows:

PRINCIPLE OF ARGUMENT 3.1 *Let $f(z)$ be a complex-valued function meromorphic in a simply connected region of the complex plane and γ a positively oriented Jordan curve not passing through any zero or pole of $f(z)$, then the number of zeros N and the number of poles P (counting multiplicities) of $f(z)$ in the interior of γ satisfy:*

$$N - P = \frac{1}{2\pi I} \oint_{\gamma} \frac{\frac{d}{dz} f(z)}{f(z)} dz \tag{1}$$

where $I^2 = -1$.

If the function $f(z)$ is analytic within the respective region, the principle of argument can be used to determine the number of zeros of $f(z)$ inside the curve γ. Even if $f(z)$ has some poles, and if we can write $f(z) = g(z) * h(z)$ where $g(z)$ does not have zeros and $h(z)$ does not have any poles, then the zeros of $f(z)$ are precisely those of $h(z)$, and the principle of argument can be used in this case for $h(z)$ instead of $f(z)$.

The principle of argument is not as elegant as it may seem, since the integrand in (1) is the logarithmic derivative of $f(z)$; the anti-derivative of which is the $\log(f(z))$, which is a multi-valued function in the field of complex numbers. Therefore, unless the considered function is relatively simple, the integral in (1) cannot be computed in closed form. Since we are to give an algorithm, that is, an automated procedure, for finding the zeros of $f(z)$, we can by no means assume that the function $f(z)$ (which is the input of our algorithm) is simple. We, therefore, shall have to resort to a numerical approximation of the above integral.

Using a quadrature scheme to evaluate the above integral will have a great impact on the permissible domain of functions in our context. We only are allowed to assume that the function $f(z)$ is analytic. However, this is not sufficient to get useful results using quadrature methods. The given function $f(z)$ is not only to be analytic but for practical reasons we also require that its logarithmic derivative is treatable by the used quadrature method. More formally, let

$$g(z) = f'(z)/f(z)$$

where $I(g)$ is the exact value of the integral in (1), and $R(g)$ is an approximated value achieved using the quadrature rule R, then the error, $E(g)$, made in using R to approximate $I(g)$ is:

$$E(g) = I(g) - R(g) \tag{2}$$

Thus, for a given quadrature rule R, the quadrature algorithm will only be useful if $E(g)$ is small enough. Hence, the algorithm to be presented in the next section will give good results if and only if $f(z)$ is an analytic function such that $E(g)$ is small enough. The fact that the sought value of the integral is an integer, makes extremely precise evaluation of $I(g)$ unnecessary. Even better, as we alluded to earlier, the algorithm we are going to present is targeted to functions where the number of zeros $I(g)$ is expected to be a large one, and we are often interested (if at all) in locating only some of the zeros. Thus, the approximation rule will be sufficient in our case for more problem instances than is usually the case[1].

[1]Often we will be interested only in knowing whether $I(g)$ is 0 or not.

Now, let us turn our attention to the main idea of the algorithm to be presented in Section 4. In more general terms, the problem statement could be rephrased as follows:

Given a function f analytic in a domain D and a bounded connected subdomain D' of D that does not have zeros of f on its boundary, find those zeros of f that lie inside D'. In principle, D' does not need to be part of the real axis (but this the case we are interested in). To find the zeros lying in D', we make use of what we call a *(zero) sweeping function* w. The function w should be analytic and reversible in D. Hence, we require that w^{-1} be defined throughout D. This calls for using a simple transformation like a (bi)linear one as a sweeping function. However, there is no need to restrict w to this kind of functions. The rationale behind the introduction of such a function w is to be able to "filter" the zeros of interest from the (perhaps many) zeros existing nearby. (The exact meaning of this filtering will be clear as we proceed). Analytically, we will treat the transformed function $f(w(z))$ instead of $f(z)$. We must however make sure that:

(a) The zeros of $f(w(z))$ are easier to approximate that those of $f(z)$.

(b) $w(z)$ and $w^{-1}(z)$ are easily computable for any z in the corresponding domain.

The reader should keep in mind that we introduced the sweeping function w with the only aim of separating the zeros we are interested in from the rest of the zeros. We emphasize that the function w is not related to any quadrature method we shall use to pinpoint the zeros, for example, by virtue of the principle of argument. Hence, the function w is inherently related the geometry of the zeros of f but has no direct relation to the evaluation process of these zeros. The function w is not just a substitutive transformation for integrals, but on the other hand, it is a general consequence of the requirements (a) and (b) that integrative numerical evaluation is more reliable using such a function w.

Having stated the main requirements and introduced the sweeping function, we should now give some simple examples prior to discussing the general algorithm. Suppose we are given the function $f(z) = (z-1)(z-(1+\epsilon))g(z)$ for small $\epsilon > 0$. Since we know that $f(z)$ has at least two zeros which are very close to each other, the numerical separation of the zeros of $f(z)$ would be a very instable process. Instead, we could define the sweeping function $w(z) = z/M$ with $M * \epsilon >> 1$ and solve the problem $f(w(z)) = 0$. The problem $f(w(z)) = 0$ is much more stable than the problem $f(z) = 0$. We could have used $w(z) = \log(z)$ in this example which would also lead to a more stable process. However, if the desired zeros are complex, the function $\log(z)$ is to be used with usual restrictions because of its discontinuity.

Sometimes we may want to use multiplicative sweeping functions. If we know that the function f has well-separated zeros in a large connected domain, it is worthwhile to down-size the domain for practical reasons of numerical integrability. This can be done via sweeping functions of the form $w(z) = M * z$ or $w(z) = exp(z)$. However, we emphasize again that the decision about which sweeping function to use is related to the geometry of the zeros of f and needs thorough analysis in general. We cannot expect that this analysis be automated for a large variety of functions and/or domains. The case when $f(z)$ is a polynomial is of primary interest in applications. Our method is unique in handling polynomials in the sense that the algorithm's complexity is not dependent on the degree of the polynomial. This is best understood by means of the following important definition.

DEFINITION 3.1 *Let a polynomial $p(z)$ with degree n be given in its Taylor representation*

$$p(z) = \sum_{i=0}^{n} a_i x^i$$

We define the size of the $p(z)$ as follows:

$$s(p) = |\{a_i : a_i \neq 0\}|$$

Informally, $s(p)$ counts the number of terms in the $p(z)$. It is important to realize that the following simple fact always holds:

$$s(p) <= n + 1$$

Moreover we easily can develop polynomials with $s(p) << n$. Or in other words, cases where n is very large are never combined (for practical reasons) with a large $s(p)$. This is very plausible if you think that n may be $\Omega(2^k)$ for some parameter k. Because we somehow must mechanically (or manually) generate all terms of $p(z)$, we are forced to keep $s(p)$ in $O(q(k'))$ for some polynomial q and parameter k', even if n is of very high order, in order to be able to work properly with $p(z)$. Making an algorithm's complexity depend on $s(p)$ instead of n is thus a big advantage unless n is small. The reader should be aware of the fact that zero-finding algorithms for polynomials in general work in $O(q(n))$ where $q(n)$ is some polynomial in the degree n (of $p(z)$). Only the Descartes rule of signs is an exception in this regard is as much as it can decide about the existence of zeros in the given interval in $\Omega(s(p))$ steps instead of $O(q(n))$ steps. However, even in this case if the number of sign changes is even (and not zero) the performance is deteriorated by (needed) additional bisections.

The algorithm presented in this paper is based on the measure $s(p)$ instead of n if $f(z)$ is a polynomial. Thus, in principle, the algorithm is able to work with high-degree polynomials as easily as with low-degree polynomials. However, we still have the problem of the accuracy of produced results if n is very large, since in this case the polynomial may oscillate too much in the considered interval.

4. The Algorithm

The algorithm presented in this section will focus on the filtering of the *real zeros* of a given analytic function. We will tackle the difficult case of functions having dense complex zeros around the real axis. Also, as stated earlier we do not assume that the number of zeros is small. Thus, we will be in general interested in locating some of the real zeros or merely in deciding whether or not the function has at least one real zero in the a given (real) interval $[a, b]$.

The main idea of the algorithm is as follows. Suppose we want to locate some real zeros of f and we know that any non-real zero z has the property that is, $Im(z) \geq \epsilon$. (This property means that the non-real zeros are not closer than ϵ to the real axis). Then, we define the sweeping function $w(z) = z/M$ with $M \geq 1/\epsilon$. In this way, the function $f(w(z))$ will have the same real zeros of $f(z)$ scaled up by the factor M in the interval $[M * a, M * b]$. Moreover, the function $w(z)$ has the properties (a) and (b) mentioned in Section 3.

Before investigating the approximation of the number of zeros, we should clarify one issue related to the use of the algorithm as an automated routine. We will require from the algorithm to give us a good result concerning the real zeros and their number for any given accuracy. This shall be achieved by making the value of ϵ a parameter of the algorithm that the user is able to specify. In so doing, the algorithm is effectively able to return the correct result to any desired accuracy. Clearly, the parameter ϵ will be in general determined after some analysis of the function under study. However, in the worst case, if namely the geometry of the zeros of the function f is not reliably known, the user can still use a number close to (but greater than) the smallest representable positive machine number if high accuracy is important. In a symbolic environment, the user is able to choose ϵ many orders of magnitudes smaller than typical machine accuracies. This last method though theoretically advisable might induce high numerical instability in practice. Actually, a good value of ϵ is in general not easily found.

The preceding considerations about $f(z)$ and the desired accuracy ϵ already partially determine the input of our algorithm. Thus given a function and an accuracy ϵ, the first steps of the algorithm is to perform the transformations:

(1) $M := 1/\epsilon$;

(2) $w(z) := z/M$;

(3) $[a, b] := [a * M, b * M]$;

(4) $f(z) := f(w(z))$.

We now focus on how to approximate the number of the real zeros in the *new* interval $[a, b]$. We will make use of the principle of the argument (see Section 3) with regard to the new function $f(z)$, that is, instead of (1), we will be using the following formula:

$$N = \frac{1}{2\pi I} \oint_{\gamma} \frac{f'(w(z))(\frac{d}{dz}w(z))}{f(w(z))} dz \tag{3}$$

However, we have to determine first a closed contour around the interval $[a, b]$. For reasons of accuracy, a circular contour is the best choice. Periodicity of the integrand calls for the use of the trapezium rule as a quadrature rule for the approximation of the integral. This rule is extremely reliable and of course (as usual) very efficient in case of periodicity. The reason lies in the well-known Euler-Maclaurin expansion of the trapezium rule. Unfortunately, we would be faced with a problem if we use a circular contour. The problem is best explained using Figure 1. We notice that after the above 4 steps of the algorithm have been used, the real zeros of $f(z)$ will move horizontally by the factor M. At the same time, the non-real zeros move horizontally as well as vertically again by the factor M. If we choose now a circular contour that includes all potential real zeros of $f(z)$, we cannot make sure that all non-real zeros are outside the domain within the contour. Hence, this solution is not feasible unless the factor M is not too large (that is non-real zeros are not too close to the real axis) in which case we may be able to decompose the contour into smaller (but not too many) adjacent circles and perform the numerical quadrature on each of them. However, this would contradict the main assumption of our problem, namely, that M is large in general.

Instead of a circular contour we will be using a rectangular one. A rectangular contour, though its numerical quadrature treatment is cumbersome and often less reliable, is the only halfway acceptable contour in our case. With careful shaping we are able to use a rectangular contour to achieve a desired accuracy. As indicated in Figure 2, the rectangular contour can be shaped so that:

(a) No potential real zeros of $f(z)$ lie outside the contour.

(b) No potential non-real zeros of $f(z)$ lie inside the contour.

Properties (a) and (b) clearly explain why we spoke of "filtering" the desired zeros by means of a sweeping function earlier in this paper. We can achieve the desired properties (a) and (b) by choosing B (see Fig. 2) in such a way that it is large enough as to avoid integration near the real axis and small enough as to avoid non-real singularities of the logarithmic derivative present under the integral sign in formula (3).

The next issue is to determine a good value of B. Obviously B must not be very small since then the contour would be close to potential real-valued singularities of the logarithmic derivative. Our approach for choosing B is preventive. When we fix the value of M, we tend to be very restrictive. That is, when the user specifies $\epsilon = 1/M$, what the algorithm will actually do is to set:

$$M := M + B \tag{4}$$

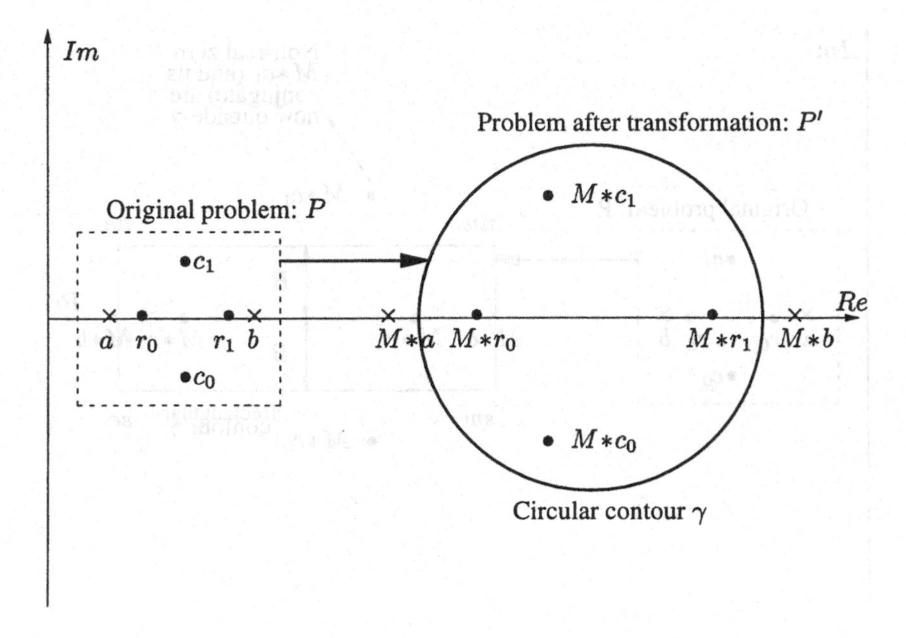

Figure 1. After transforming the problem via $w(z)$, some non-real zeros (here $M * c_0$ and $M * c_1$) may unavoidably remain in any circular contour containing all zeros in [M*a, M*b] (here $M * r_0$ and $M * r_1$).

The algorithm will therefore effectively work with $M + B$ instead of M. This has the nice side effect that we are now able to choose B freely from within the algorithm. You can, therefore, think of B as a configuration parameter of the algorithm itself. In any instance, B should be chosen relatively large compared to 1. In our experiments, the value of B was chosen to be in the range [1, 100]. However, larger values are acceptable too and may lead to better results when used. This method for the determination of B will not work if the user chooses ϵ to be very close to the smallest representable positive number in the machine. This case is deemed unusual in this treatment and should be avoided/prevented for reasons of numerical stability.

We still did not specify the quadrature rule to be used. We already mentioned that the trapezium rule is unsurpassed for circular contours. In our case, though the contour is rectangular, we decided to use the trapezium rule for reasons of efficiency. As should any automatic quadrature procedure, our quadrature algorithm is adaptive. Adaptivity is needed since we cannot expect the user to know a priori a minimum number of function evaluations needed to achieve the desired accuracy. Hence, we use an adaptive trapezium quadrature, with accuracy requirement *eps* in the range [0.1, 0.4]. This is sufficient to distinguish two consecutive integers. It is worth mentioning in this context that

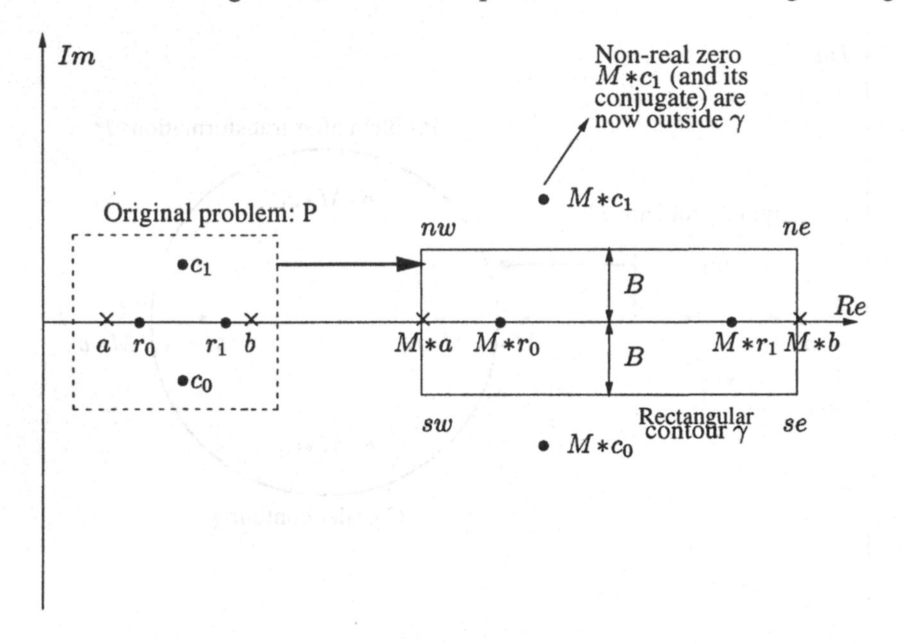

Figure 2. The rectangle contour γ with length $M * (b - a)$, width $2 * B$, and edges (sw, se, ne, and nw) encloses now only the real zeros of $f(z)$ scaled up by the factor M.

there is no need to let the user specify *eps*, since the algorithm is intended to approximate an integer-valued integral and not to be used as a general-purpose adaptive quadrature method. As with any adaptive method, the quadrature procedure relies on a subdivision of the treated interval that depends on the characteristics of the integrand in different sections of the interval. The implemented algorithm is therefore recursive. The adaptive trapezium quadrature procedure stops if either of the following conditions occurs:
(a) The required accuracy is reached in the respective sub-interval.
(b) The size of the treated subinterval (which is halved in each recursive call) is less than a pre-assigned threshold.

The threshold mentioned above is also configuration parameter of the algorithm and should be set as small as possible. It is also possible to consider this threshold as an input parameter of the algorithm, in which case the user is responsible for its determination. Obviously, only case (a) of above would lead to the desired result. However, in practice useful results can also be achieved in spite of the termination s enforcing criterion (b), if the threshold is chosen carefully.

Now we address one of the most crucial issues in the evaluation process of the integral. This issue deals with the problem that the integration interval

$[a, b]$ is per construction very large. It goes without saying that the error in the evaluation of the integral (see (2)) is among others super-proportional to the size of the interval[2]. In the adaptive version of the trapezium rule, this dependency will manifest itself in an almost non-terminating computation; a highly undesirable situation. To avoid this problem, we resort to the classical substitution technique. There are many ways to perform an interval-shrinking substitution and we decided to use the logarithm function. Thus, the integral in (3) is further transformed to:

$$N = \frac{1}{2\pi I} \oint_{\ln(\gamma)} \frac{f'(w(e^z))w'(e^z)e^z}{f(w(e^z))} dz \tag{5}$$

Last integral is more precisely the sum of four integrals along the sides of the rectangle[3]:

$$N_i = \frac{1}{2\pi I} \oint_{\ln(z_i)}^{\ln(z_{i+1})} \frac{f'(w(e^z))w'(e^z)e^z}{f(w(e^z))} dz \tag{6}$$

where $(z_i, z_{i+1}) \in \{(sw, se), (se, ne), (ne, nw), (nw, sw)\}, i \in \{1, 2, 3, 4\})$.
Thus, N of (5) is then (clearly, the $N_i's$ need not to be integers or even real):

$$N = N_1 + N_2 + N_3 + N_4$$

It is pertinent to point out here that this last integral transformation of equation (6) is a typical one and is different from the transformation made in (3). There, we needed to know the geometry of zeros in order to find a convenient function $w(z)$. However, the transformation here does not need any information about the integrand's zeros. In performing the transformation (6) we have to make sure that we use the correct branch of the logarithm for the different four integrals. However, for different integrals different branches can be used. The price paid in issuing the transformation in (6) is:

(1) The integrand is now more complex increasing the time for function evaluations.

(2) The integrand might lose some desirable properties after the transformation (less smooth, non-periodic, etc.).

(3) The second derivative of the integrand may highly increase in absolute values contra-acting the shrinkage of the integration interval in the error term.

---

[2] Precisely, the error of the m−point trapezium rule is $-((b-a)^3/12m^2)f''(\xi)$ for some ξ in (a, b).
[3] Thus the following quadrature discussion equally apply to all four integrals.

Using (6) the algorithm is able to unambiguously estimate the number of the zeros $N(a, b)$ in the interval $[a, b]$. If this number is zero, the algorithm terminates. If $N(a, b)$ is small, say $N(a, b) \leq 20$, the algorithm is reused recursively for the subintervals $[a, (a+b)/2]$ or $[(a+b)/2, b]$ depending on which subinterval includes more zeros (counting multiplicities). This bisection is performed as long as $N(a, b)$ is different from zero. In order to determine a zero with a specified accuracy ϵ_0, the algorithm uses this bisection method until $b-a \leq \epsilon_0$. Here, ϵ_0 is a user-supplied input of the algorithm. In this way, if the original interval [a, b] only includes a small number of zeros, those zeros are determined to any desired accuracy. The complexity of the bisection scheme for an interval [a, b] is $O(log(b-a))$. As mentioned in the Section 3, one of the strengths of the algorithm is that it can deal with functions having a large number of zeros as easily as with those with small number of zeros.

Let us shed some light on the performance and the numerical reliability of the algorithm. The numerical reliability of the algorithm is only limited by the convenience of the quadrature rule. Taking into consideration that the number we are seeking is an integer, we see that even a crude approximation is highly appreciated. There is an "apparently" good way of enhancing the numerical reliability of the quadrature process by using $f^2(z)$ instead of $f(z)$, thus halving the accuracy demands, since the new zeros are double zeros, and we only need to know $N(a, b)$ to 0.9 accuracy (say), since $N(a, b)$ is known to be an even number in this case. Unfortunately, this will not work, since formula (6) includes the logarithmic derivative making any exponentiation behave like a mere multiplication by a constant.

The decision for using the trapezium rule has also to do with performance. It is well known that in the trapezium rule pre-computed functions values need not be computed again if the integration interval is further refined. Gauss quadrature is extremely time consuming in this respect, since already computed function values are not reused. Moreover, if the function $f(z)$ is periodic and the size of the interval $[a, b]$ is a multiple of its period, we know that the trapezium rule is extremely reliable. Since the algorithm s quadrature process is adaptive, the number of needed function evaluations will in general depend on the function $f(z)$. The stopping criterion (b) (artificially) integrated in this process is only a mean to limit this number even if some accuracy is lost. The bisection procedure, however, is extremely fast $(O(log(b-a))$. Thus pinpointing a zero is done in logarithmic time (for example, versus binary search).

In calculating the values of the four integrals in (6), we are occasionally able to make the following optimizations:

1 If the function $f(z)$ is real on the real axis, it can be easily shown that the zeros of $f(z)$ will be either real or conjugate pairs. It is therefore needless to evaluate both integrals N_1 and N_3 of formula (6). We only

need to compute one integral, since their values will be conjugate:

$$N_1 + N_3 = 2 * Re(N_1) = 2 * Re(N_3)$$

2 If $f(z)$ is periodic and the size of the interval $[a, b]$ is a multiple of its period, there is no need to compute the integrals N_2 and N_4 at all, since their values will cancel. Thus, we see that periodicity of the integrand both lowers the reliability demands and enhances performance (a rare combination).

3 If $f(z)$ satisfies both above conditions, we merely need to compute one integral in formula (6), namely, N_1 or N_3.

Another minor optimization that would enhance performance is to only evaluate the real parts of the integrals in (6), that is:

$$N(a, b) = Re(N_1) + Re(N_2) + Re(N_3) + Re(N_4)$$

5. Examples and Experimental Results

We implemented the described algorithm in Maple 8.01. The actual implementation is straightforward in the sense that we did not take any precaution related to round-off errors. The results are promising though. Below, we give an account of some experiments with the implemented algorithm.

We have selected three types of functions: polynomials, trigonometric functions, and other more involved functions containing transcendental ones. If applicable, we indicate the source of the selected function. Also, in each case we indicate the respective interval $[a, b]$ and the values of the parameters M and B. These values are important since they reflect the needed accuracies of the user.

Polynomials

$$f_1(z)(z-4)(z-5)^3(z-4-\frac{1}{1000}I)(z-4+\frac{1}{1000}I)$$

This polynomial has 4 real zeros (counting multiplicities) and two conjugate complex zeros:

$$z_0 = 4,\ \ z_1 = 5,\ \ z_3 = 4 + \frac{1}{1000}I, \overline{z_3}$$

Notice that the conjugate pair is very close to the real zero z_0.

Applying the algorithm with $[a, b] = [3, 3 + 2\pi], M = 10000$, $B = 5$. We obtained the result of $4.0000416415776037389 + 0.I$. This is the expected result with enough accuracy.

We need to mention here that in the actual implementation it was not possible to work with complex zeros much closer than 10^{-3} to the real axis. This problem needs further investigation as pointed out in the conclusion.

The next polynomial is an extremely ill-conditioned one of (Delves and Lyness, 1967(a)):

$$\begin{aligned} f_2(z) \;=\; & 1250162561\, z^{16} + 385455882\, z^{15} + 845947696\, z^{14} \\ & +240775148\, z^{13} + 247926664\, z^{12} + 41018752\, z^{10} \\ & +9490840\, z^{9} + 4178260\, z^{8} + 837860\, z^{7} + 267232\, z^{6} \\ & +44184\, z^{5} + 10416\, z^{4} + 1288\, z^{3} + 224\, z^{2} + 16\, z + 2 \end{aligned}$$

Using the same parameters as in the preceding example, the algorithm gives the result of $-0.13505417 + 106{-}5 + 0.I$. This is also the expected result, since this polynomial does not have any real zero.

The result of the next polynomial shows one of the strongest features of the algorithm. The polynomial has a very large degree and we anticipate knowing the number of its real zeros in the mentioned interval. The polynomial is:

$$f_3(z) = (z-4)(z-5)^3(z-4-\frac{1}{10}I^{10000}(z-4+\frac{1}{10}I)^{100000}$$

We used the algorithm for this polynomial with $M = 20$, $B = 1$. The result obtained is $3.992583 - 0.I$. This is an extremely good result, since out of 200004 zeros of $f_3(z)$ only 4 (counting multiplicities) are real. Notice that the 200000 complex zeros are close to real axis.

Trigonometric Functions

It is often beneficial to make the use of the periodicity of this type of functions. The following function is a simple trigonometric polynomial:

$$f_4(z) = cos(5z) - 2cos(2z) + cos(z) + 3$$

We obtained the correct result (*i.e.*, 2) with our algorithm:

$$2.001582398 + 0.636619772210^{-16} I$$

We used here:

$$[a, b] = [1, 2\pi + 1], M = 100000, B = 100$$

The next function is based on the polynomial $f_2(z)$

$$f_5(z) = f_2(cos(z) + 1) - 4$$

Maple shows that this (periodic) function has two simple real zeros (close to 3) in the interval $[0, 2\pi]$. Our algorithm's result confirms that:

$$1.997902195 + 0.359690171310^{-15} I$$

We used here the same values of a, b, M, and B as in the preceding example.

Other Functions

The next function is a mixed sine/exponential function.

$$f_6(z) = e^{(\frac{z}{4}-1)} + 10sin(\frac{z}{2} - 5)$$

Applying our algorithm with same the parameters as in the preceding example, we obtained the result of $1.006411448 - 0.186211283410^{-16} I$. This confirms the Maple plot function, which shows a single real root in the corresponding region.

The next function is used in practice to model heat propagation (Greenleaf, 1972) and one is interested in general in some of its (many) real zeros. Here, we used the instance:

$$f_7(z) = tan(z) - z$$

With $B = 20$ and without changing the other parameters, we obtained the result of $4.415547706+0.111408460110^{-17} I$. This is still an acceptable result for the true value 4.

The final example is based on the Bessel function. The function is often used in physical applications to model wave reflection (Kravanja, Barel and Van, 2000).

$$f_8(n, z) = J_n(z - 2\pi) + I J_{n+1}(z - 2\pi)$$

It is known that if $n > 0$, $f_8(n, z)$ admits only one real zero multiplicity n at 2π with all remaining zeros non-real. We anticipate verifying this theoretical result using our algorithm for n in the range [1, 10]. The following table lists the results for the treated values of n.

N	Obtained Result
1	$1.321014932 + 0.0002922914206I$
2	$2.253817259 + 0.0002022027040I$
3	$3.208294409 + 0.0001329376255I$
4	$4.175709689 + 0.0000884997349I$
5	$5.151328999 + 0.0000605998921I$
6	$6.132658423 + 0.0000427817031I$
7	$7.117892159 + 0.0000310835358I$
8	$8.105992370 + 0.0000231751786I$
9	$9.096218075 + 0.0000176776976I$
10	$10.08803592 + 0.0000137573736I$

The algorithm's parameters in all these experiments were set as follows:

$$[a, b] = [2, 2\pi + 2], \quad M = 1000, \quad B = 10$$

For all values of n in [1, 10] we have acceptable results. One can also easily see that the experiments with the values 100 and 1000 for n and we obtained respectively:

$$100.0099893 + 0.2195117 \ 10^{-7} I$$

$$1000.001071 + 0.36733 \ 10^{-10} I$$

The sweeping function w that has been presented separates the set of zeros that the user is interested in from the remaining set of zeros. In fact, w is a filter as we assume the number of zeros is large. The algorithm, which is based on finding a rectangular contour in the interval $[a, b]$ is amenable to parallelism and can be implemented on a parallel computer. At each step, the rectangular contour is subdivided into sub-rectangles. The sub-rectangles being found to contain zeros are further subdivided. If a sub-rectangle does not contain a zero it is discarded. If we continue this subdivision process we will likely pinpoint the real zeroes. Each sub-rectangle generates a set of new sub-rectangles. Each sub-rectangle being considered for the next iteration should have the following properties: (a) no potential real zeros are outside the contour, (b) no potential non-real zeros are inside the contour. In addition to the configuration parameter B and the factor M, we can add other parameters (*i.e.*, efficiency, speedup, accuracy, reliability) that may influence the results positively. The number of processors available increases the number of subdivisions, however, the algorithm may not give the efficiency that would be expected. Our initial analysis indicates that it is likely possible to speed up the sequential algorithm that has been presented. This work is being investigated under these lines and will be our forthcoming research paper.

6. Conclusion

The main achievements of this work are:

(a) Addressing the problem of filtering interesting zeros of a given analytic function. (b) Introducing the concept of a sweeping function as a tool for such filtering. (c) Applying the concept to the selection of real zeros of an analytic function. (d) Constructively demonstrating the usefulness of the introduced concepts by means of an algorithm that proved to be able to handle a large type of analytic functions. (e) Assuring that the algorithm is able to handle extreme cases like polynomials with high degrees, analytic functions with large number of zeros, and zeros that are very close together.

There is no doubt that the described algorithm even in its current form will provide analysts with much-needed insights for their functions. However, in the design of the algorithm we left one important item unanswered, namely, how can the user/analyst be guided in the process of finding good values of the prescribed accuracy (*i.e.,* M). This question is still open. Another more important question in this context is to study the influence of the parameters M and B (and other figuration parameters) on the accuracy of the algorithm's results. This may lead to dependencies on the used quadrature rule. It may outcome that a Gauss-like quadrature rule is needed to enhance independency of the algorithm's accuracy on the parameters M and B. All in all, we need more analysis and experiments in this respect and we need to know more about the overall conditioning of the algorithm with respect to its parameters.

We also introduced the concept of sweeping functions. We, however, did not provide any hint selecting a good such function. This problem is also open for the time being and needs more clarification. A related problem is how to find sweeping function for general domains in the complex plane. Recall that the present paper only dealt with domains of the real axis (real intervals).

Acknowledgments

We would like to thank the referees for their valuable comments that helped improve the paper.

References

Atanassova L. On the Simultaneous Determination of the Zeros of Analytic an Function Inside a Simple Smooth Closed Contour in the Complex Plane. *J. Comput. Appl. Math.*, 50:99–107, 1994.

Delves L. M., Lyness J. N. A Numerical Method for Locating the Zeros of Analytic Functions. *Math. of Comput.*, 21(100):543–560, 1967.

Delves L. M., Lyness J. N. On Numerical Contour Integration Round a Closed Contour. *Mathematics of Comput.*, 21(100):561–577, 1967.

Douglas J., Burden R. Numerical Methods. Thomson Books/Cole, 2003.

Drucker D. S. A Second Look at Descartes Rule of Signs. *Mathematics Magazine*, 52: 237–238, 1979.

Foster L. V. Generalizations of Laguerre's Method: Higher Order Methods. *SIAM Journal on Num. Analysis*, 18(6):1004–1018, 1981.

Greenleaf F. P. Complex Variables. W. B. Saunders Company, 1972.

Henrici P. Applied and Computation Complex Analysis. *Power Series- Integration-Conformal Mapping-Location of Zeros, Vol 1*. Wiley, 1974.

Herlocker J., Ely J. An Automated and Guaranteed Determination of the Number of Roots of Analytic Functions Interior to a Simple Closed Curve in the Complex Plane, *Reliable Comput.*, 3:239–250, 1995.

Kravanja, P., Barel, M. Van. Computing the Zeros of Analytic Functions, Series. *Lecture Notes in Mathematics, VII*, 2000.

Lehmer D. H. A Machine Method for Solving Polynomial Equations. *J. Assoc. Comput. Mach.*, 8:151–162, 1962.

Li T. On Locating All Zeros of an Analytic Function within a Bounded Domain by a Revised Delves/Lyness Method. *SIAM Journal on Numerical Analysis*, 20(4):865–871, 1983.

Olver H. W. The Evaluation of Zeros of High-Degree Polynomials. *Ph. Trans. of the Royal Soc. of London, (A)*, 244(885):385–415, 1952.

Petcovic M. S. et. al. Weierstrass Formula and Zero-Finding Methods. em Numerische Mathematik, 29: 353–372, 1995.

Petcovic M. S., Marjanovic Z. M. A Class of Simultaneous Methods for the Zeros of Analytic Functions. *Comput Appl. Math.*, 22(10): 79–87, 1991.

Vincent A. J. H. Sur la Resolution des Equations Numeriques. *Journal de Mathematiques Pures et Appliquees*, 341–371.

Wilf H. S. A Global Bisection Scheme for Computing the Zeros of Polynomials in the Complex Plane. *Journal of the Association for Computing Machinery*, 25(3): 415–420, 1978.

Ying X., Katz I. N. A Reliable Argument Principle Algorithm to Find the Number of Zeros of an Analytic Function in a Bounded Domain. *Numerische Mathematik*, 53:143–163, 1988.

REDUCING OVERHEAD IN SPARSE HYPERMATRIX CHOLESKY FACTORIZATION

Jose R. Herrero and Juan J. Navarro
Computer Architecture Department, Universitat Politecnica de Catalunya *
Jordi Girona 1-3, Modul D6, E-08034 Barcelona, Spain
{josepr,juanjo}@ac.upc.es

Abstract

The sparse hypermatrix storage scheme produces a recursive 2D partitioning of a sparse matrix. Data subblocks are stored as dense matrices. Since we are dealing with sparse matrices some zeros can be stored in those dense blocks. The overhead introduced by the operations on zeros can become really large and considerably degrade performance. In this paper, we present several techniques for reducing the operations on zeros in a sparse hypermatrix Cholesky factorization. By associating a bit to each column within a data submatrix we create a bit vector. We can avoid computations when the bitwise AND of their bit vectors is null. By keeping information about the actual space within a data submatrix which stores non-zeros (dense window) we can reduce both storage and computation.

Keywords: Sparse Hypermatrix Cholesky, bit vector, dense window

1. Introduction

Sparse Cholesky factorization is heavily used in several application domains, including finite-element and linear programming methods. It often takes a large part of the overall computation time incurred by those methods. Consequently, there has been great interest in improving its performance (Duff, 1982; Ng and Peyton, 1993; Rothberg, 1996). Methods have moved from column-oriented approaches into panel or block-oriented approaches. The former use level 1 BLAS while the latter have level 3 BLAS as computational kernels (Rothberg, 1996). Operations are thus performed on blocks (submatrices). A matrix M is divided into submatrices of arbitrary size. We call

*This work was supported by the Ministerio de Ciencia y Tecnologia of Spain and the EU FEDER funds (TIC2001-0995-C02-01)

M_{br_i,bc_j} the data submatrix in block-row br_i and block-column bc_j. Figure 1 shows 3 submatrices within a matrix. The highest cost within the Cholesky factorization process comes from the multiplication of data submatrices. In order to ease the explanation we will refer to the three matrices involved in a product as A, B and C. For block-rows br_1 and br_2 (with $br_1 < br_2$), and block-column bc_j each of these blocks is $A \equiv M_{br_2,bc_j}$, $B \equiv M_{br_1,bc_j}$ and $C \equiv M_{br_2,br_1}$. Thus, the operation performed is $C = C - A \times B^t$, which means that submatrices A and B are used to produce an update on submatrix C.

Block size can be chosen either statically (fixed) or dynamically. In the former case, the matrix partition does not take into account the structure of the sparse matrix. In the latter case, information from the *elimination tree* (Liu, 1990) is used. Columns having similar structure are taken as a group. These column groups are called *supernodes* (Liu et al., 1993). Some supernodes may be too large to fit in cache and it is advisable to split them into *panels* (Ng and Peyton, 1993; Rothberg and Gupta, 1991). In other cases, supernodes can be too small to yield good performance. This is the case of supernodes with only one or a few columns. Level 1 BLAS routines are used in this case and the performance obtained is therefore poor. This problem can be lightened by *amalgamating* several supernodes into a single larger one (Ashcraft and Grimes, 1989). Then, some null elements are both stored and used for computation. However, the use of level 3 BLAS routines often results in some performance improvement.

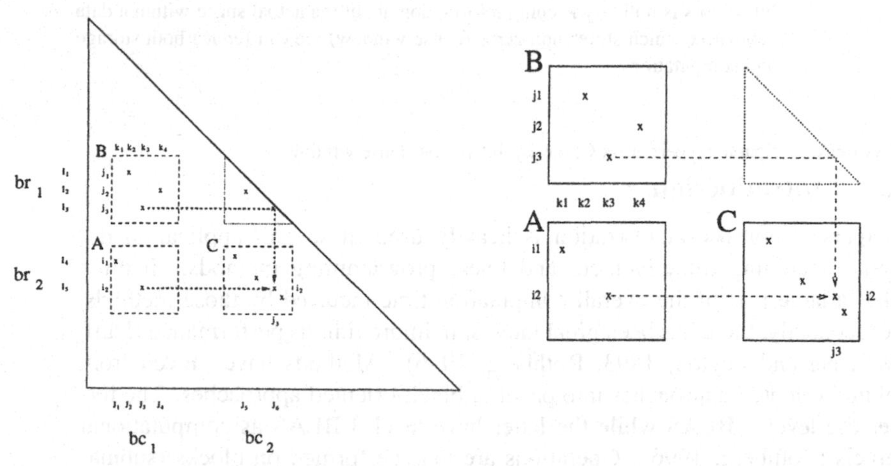

Figure 1. Blocks within a matrix: definition and example of use.

In this paper we address the problem of dealing with zeros within submatrices. This problem can arise either when a fixed partitioning is used or when supernodes are amalgamated. We present several techniques we have used to reduce the overhead introduced by the fact that data submatrices are stored and operated on as dense. This is interesting in the context where we are focused: the optimization of the *Hypermatrix* (Fuchs et al., 1972) data structure.

Hypermatrix representation of a sparse matrix

Sparse matrices are mostly composed of zeros but often have small dense blocks which have traditionally been exploited in order to improve performance (Duff, 1982). Our application uses a data structure based on a hypermatrix (HM) scheme (Fuchs et al., 1972; Noor and Voigt, 1975). The matrix is partitioned recursively into blocks of different sizes. The HM structure consists of *N* levels of submatrices. The top *N-1* levels hold pointer matrices which point to the next lower level submatrices. Only the last (bottom) level holds data matrices. Data matrices are stored as dense matrices and operated on as such. Null pointers in pointer matrices indicate that the corresponding submatrix does not have any non-zero elements and is therefore unnecessary. Figure 2 shows a sparse matrix and a simple example of corresponding hypermatrix with 2 levels of pointers.

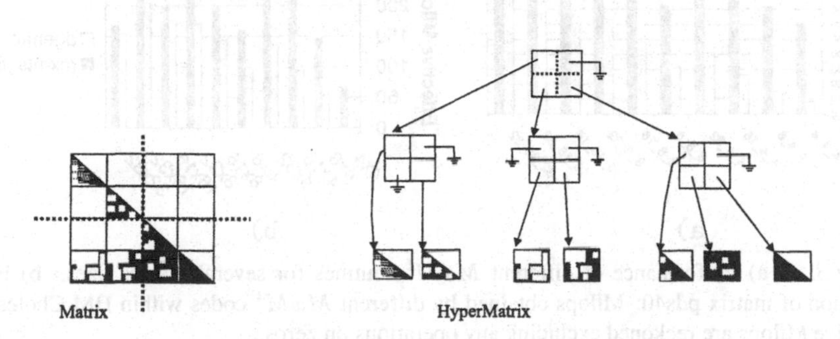

Figure 2. A sparse matrix and a corresponding hypermatrix.

The main potential advantages of a HM structure w.r.t. 1D data structures, such as the Compact Row Wise structure, are: the ease of use of multilevel blocks to adapt the computation to the underlying memory hierarchy; the operation on dense matrices; and greater opportunities for exploiting parallelism. A commercial package known as PERMAS uses the hypermatrix structure (Ast et al., 1997). It can solve very large systems out-of-core and can work in parallel. However, the disadvantages of the hypermatrix structure, namely the storage and computation on zeros, introduce a large overhead. Recently a variable size blocking was introduced to save storage and to speed the parallel execution (Ast et al., 2000). In this way the HM was adapted to the sparse matrix being factored.

Previous work

Choosing a block size for data submatrices is rather difficult. When operating on dense matrices, it is better to use large block sizes. On the other hand, the larger the block is, the more likely it is to contain zeros. Since computation with zeros is non productive, performance can be degraded. Thus, a trade-off between performance on dense matrices and operation on non-zeros must be reached. In a previous paper (Herrero and Navarro, 2003), we explained how we could reduce the block size while we improved performance. This was achieved by the use of a specialized set of routines which operate on small matrices of fixed size. By small matrices we mean matrices which fit in the first level cache. The basic idea used in producing this set of routines, which we call the Small Matrix Library (SML), is that of having dimensions and loop limits fixed at compilation time. For example, our matrix multiplication routines *mxmts_fix* clearly outperform the vendor's BLAS routine *dgemm_nts* for small matrices (figure 3a) on an R10000 processor.

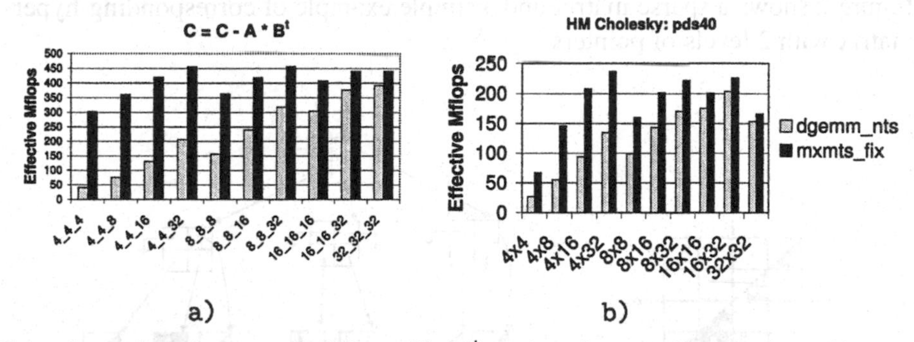

Figure 3. **a)** Performance of different MxM^t routines for several matrix sizes. **b)** Factorization of matrix pds40: Mflops obtained by different MxM^t codes within HM Cholesky. Effective Mflops are reckoned excluding any operations on zeros.

The matrix multiplication routine used affects the performance of hypermatrix Cholesky factorization. This operation takes up most of the factorization time. We found that using *mxmts_fix* a block size of 4×32 usually produced the best performance. In order to illustrate this, figure 3b shows results of the HM Cholesky factorization on an R10000 for matrix pds40 (Carolan et al., 1990). The use of a fixed dimension matrix multiplication routine speeded up our Cholesky factorization an average of 12% for our test matrix set (table 1).

Goals

The work presented in this paper focuses on the reduction of the overhead introduced by operations on zeros kept in data submatrices. In addition to the techniques mentioned above on block size reduction and the use of specialized routines, we want to reduce the amount of operations on zeros within blocks.

2. Reducing overhead

Let A and B be two off-diagonal submatrices in the same block-column. At first glance, these matrices should always be multiplied since they belong to the same block-column. However, there are cases where it is not necessary. We are storing data submatrices as dense while the actual contents of the submatrix do not necessarily have to be dense. Thus, the result of the product $A \times B^t$ can be zero. Such an operation will produce an update into some matrix C whenever there exists at least one column for which both matrices A and B have any non-zero element (e.g. column k_3 in figure 1). Otherwise, if there are no such columns, the result will be zero. Consequently, that multiplication can be skipped. In the following subsections we present several techniques which can reduce the number of non productive operations.

Bit vectors

We want to be able to avoid unnecessary matrix multiplications between matrices with elements in disjoint columns. What we need to know is whether a column within a data submatrix has any non-zero elements or not. We associate a set of bits to each data submatrix. We refer to such a set of bits as *bit vector*. Each bit in the vector is used to point to the existence of any non-zero in the corresponding column. For instance, consider matrix B in figure 4a. Let us consider column indices start at 1 (Fortran indexing). There are non-zero elements only in columns $k_2 = 3$, $k_3 = 4$ and $k_4 = 7$. Thus, only bits 3, 4 and 7 in BV_B will be different from 0. A bit-wise AND between bit vectors corresponding to matrices A and B can be used to decide whether the matrix multiplication between those matrices is necessary or not. If a single bit of the bit-wise AND results to be 1 then we need to perform the operation. If all bits are zero, then we can skip it. This test can be done in a couple of CPU cycles with an AND operation followed by a comparison to zero. The creation of the bit vectors can be done initially, when the hypermatrix structure is prepared using the symbolic factorization information. The overhead for their creation is negligible.

Dense windows within data submatrices

In order to reduce the storage and computation of zero values, we define *windows* of non-zeros within data submatrices. Figure 5a shows a window of non-zero elements within a larger block. The window of non-zero elements is defined by its top-left and bottom right corners. All zeros outside those limits are not used in the computations. Null elements within the window are still stored and computed. Storage of columns to the left of the window's leftmost column is avoided since all their elements are null. Similarly, we do not store

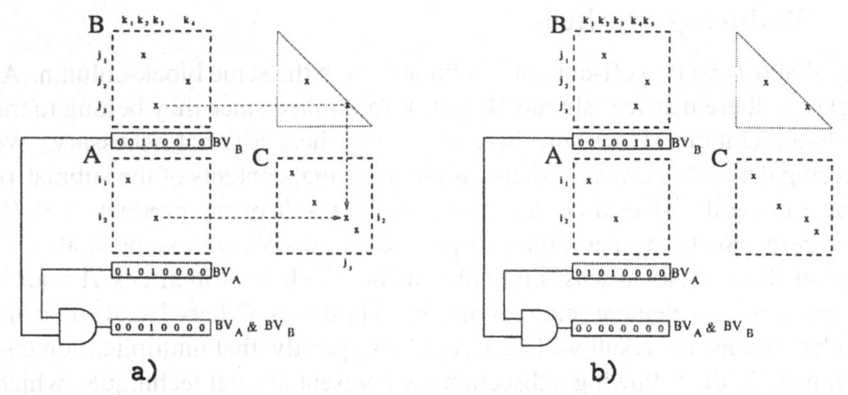

Figure 4. **a)** $BV_A \& BV_B \neq 0$: operation must be performed. **b)** $BV_A \& BV_B = 0$: operation can be avoided.

columns to the right of the window's rightmost column. However, we do store zeros over the window's upper row and/or underneath its bottom row whenever these window's boundaries are different from the data submatrix boundaries, i.e. whole data submatrix columns are stored from the leftmost to the rightmost columns in a window. We do this to have the same leading dimension for all data submatrices used in the hypermatrix. Thus, we can use our specialized SML routines which work on matrices with fixed leading dimensions. Actually, we extended our SML library with routines which have leading dimensions of matrices fixed, while loop limits can be given as parameters. Some of them have all loop limits fixed, while others have only one, or two of them fixed. Other routines have all the loop limits given as parameters. The appropriate routine is chosen at execution time depending on the windows involved in the operation. Thus, although zeros can be stored above or underneath a window, they are not used for computation. Zeros can still exist within the window but, in general, the overhead is greatly reduced.

The use of windows of non-zero elements within blocks allows for a larger default block size. When blocks are quite full operations performed on them can be rather efficient. However, in those cases where only a few non-zero elements are present in a block, or the intersection of windows results in a small area, only a subset of the total block is computed (dark areas within figure 5b).

When the column-wise intersection of windows in matrices A and B is null, we can avoid the multiplication of these two matrices (figure 6a). There are cases where the window definition we have used is not enough to avoid unnecessary operations. Consider figure 6b: there is a column-wise intersection of windows in A and B. Thus, we would perform a product using the dark area within the three matrices involved.[1]

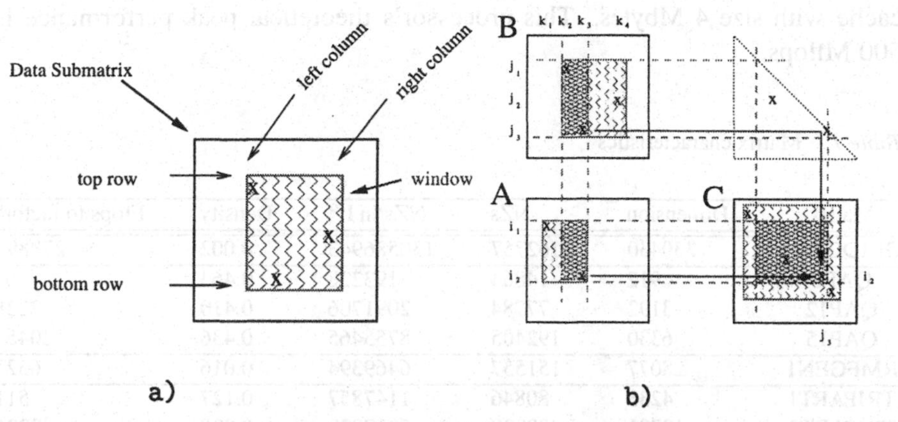

Figure 5. **a)** A data submatrix and a window within it. **b)** Windows can reduce the number of operations.

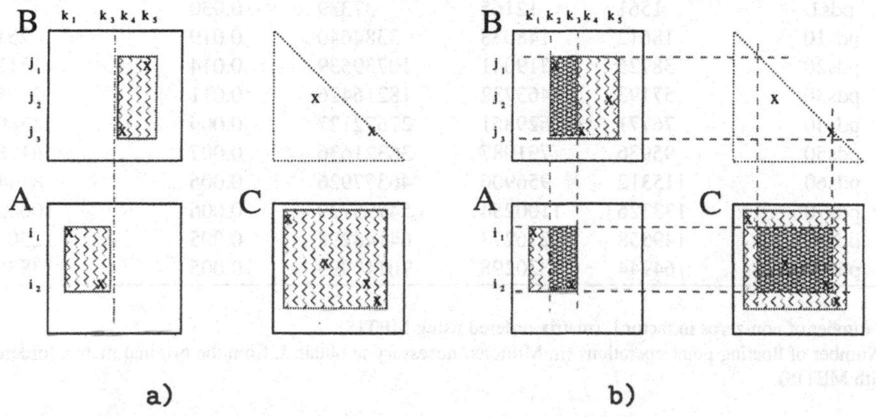

Figure 6. **a)** Disjoint windows can avoid matrix products. **b)** Windows can be ineffective to detect false intersections.

3. Results

We have used several test matrices. All of them are sparse matrices corresponding to linear programming problems. QAP matrices come from Netlib (NetLib,) while others come from a variety of linear multicommodity network flow generators: A Patient Distribution System (PDS) (Carolan et al., 1990), with instances taken from (Frangioni,); RMFGEN (Badics, 1991); GRIDGEN (Lee and Orlin, 1991); TRIPARTITE (Goldberg et al., 1998). Table 1 shows the characteristics of several matrices obtained from such linear programming problems. Matrices were ordered with METIS (Karypis and Kumar, 1995) and renumbered by an elimination tree postorder. Execution took place on a 250 Mhz MIPS R10000 Processor. The first level instruction and data caches have size 32 Kbytes. There is a secondary unified instruction/data cache with size 4 Mbytes. This processor's theoretical peak performance is 500 Mflops.

Table 1. Matrix characteristics

Matrix	Dimension	NZs	NZs in L[a]	Density	Flops to factor[b]
GRIDGEN1	330430	3162757	130586943	0.002	278891
QAP8	912	14864	193228	0.463	63
QAP12	3192	77784	2091706	0.410	2228
QAP15	6330	192405	8755465	0.436	20454
RMFGEN1	28077	151557	6469394	0.016	6323
TRIPART1	4238	80846	1147857	0.127	511
TRIPART2	19781	400229	5917820	0.030	2926
TRIPART3	38881	973881	17806642	0.023	14058
TRIPART4	56869	2407504	76805463	0.047	187168
pds1	1561	12165	37339	0.030	1
pds10	18612	148038	3384640	0.019	2519
pds20	38726	319041	10739539	0.014	13128
pds30	57193	463732	18216426	0.011	26262
pds40	76771	629851	27672127	0.009	43807
pds50	95936	791087	36321636	0.007	61180
pds60	115312	956906	46377926	0.006	81447
pds70	133326	1100254	54795729	0.006	100023
pds80	149558	1216223	64148298	0.005	125002
pds90	164944	1320298	70140993	0.005	138765

[a]Number of non-zeros in factor L (matrix ordered using METIS).
[b]Number of floating point operations (in Millions) necessary to obtain L from the original matrix (ordered with METIS).

The left half of table 2 presents results obtained by a supernodal (SN) block Cholesky factorization (Ng and Peyton, 1993). It takes as input parameters the cache size and unroll factor desired. This algorithm performs a 1D partitioning of the matrix. A supernode can be split into *panels* so that each panel fits in cache. This code has been widely used in several packages such as LIPSOL (Zhang, 1996), PCx (Czyzyk et al., 1997), IPM (Castro, 2000) or SparseM (Koenker and Ng, 2003). Although the test matrices we have used are in most cases very sparse, the number of elements per column is in some cases large enough so that a few columns fill the first level cache. Thus, a one-dimensional partition of the input matrix produces poor results. As the problem size gets larger, performance degrades heavily. We noticed that we could improve its results by specifying cache sizes larger than the actual first level cache. However, performance degrades in all cases for large matrices.

The right half of table 2 shows results obtained by several variants of our sparse hypermatrix Cholesky code. We have used SML (Herrero and Navarro,

Table 2. Supernodal vs Hypermatrix Cholesky: Mflops

	Supernodal Cholesky (Ng-Peyton)								Hypermatrix Cholesky					
Upper levels									32 x 512					
Block size									8 x 8		4 x 32			
Windows									No		No		Yes	
Bit Vectors									No	Yes	No	Yes	No	Yes
Cache	32K		512K		1M		2M							
Unrolling	4	8	4	8	4	8	4	8						
GRIDGEN1						23.8		24.4	—	—	—	—	201.2	199.5
QAP8	186.6	194.2	186.7	194.9		175.1		177.3	139.0	142.5	146.6	151.5	179.6	180.2
QAP12	118.8	102.2	181.0	223.0		215.7		166.3	160.5	176.0	174.7	197.5	246.8	247.3
QAP15	49.0	54.8	152.4	186.0		165.6		149.1	213.1	214.4	222.7	248.4	303.1	300.2
RMFGEN1	55.9	61.9	169.8	189.0		256.2		154.9	221.0	220.8	202.8	210.7	298.4	300.9
TRIPART1	118.7	176.4	175.6	177.1		160.5		164.5	151.2	170.7	151.0	152.8	203.6	207.1
TRIPART2	205.2	182.5	208.4	213.1		216.9		171.8	175.5	202.5	156.8	178.3	232.5	235.3
TRIPART3		116.9		142.7		188.2		151.3	199.0	213.1	181.7	185.3	256.6	261.1
TRIPART4		46.4		119.3	121.1	133.8		118.7	222.8	231.5	222.5	241.4	295.5	295.2
pds1	87.5	89.6		75.7		73.8			19.0	20.2	13.7	14.3	20.2	20.2
pds10	121.5	125.5		183.5		132.2			102.4	106.5	111.6	121.3	193.3	192.3
pds20	106.8	82.7		130.3		104.2			126.1	127.1	139.3	149.9	229.7	227.6
pds30	104.0	78.2		133.5		135.5			142.9	141.5	169.3	178.0	241.7	241.1
pds40	99.9	97.3		126.4		159.8			144.5	144.6	169.8	176.8	247.9	242.1
pds50	84.8	92.0		121.1		121.4			140.2	147.5	181.3	191.4	252.4	252.2
pds60	85.0	66.0		131.8		112.0			152.8	153.4	181.4	188.1	253.9	254.5
pds70		91.4		127.4		111.3			154.6	154.8	186.0	194.4	253.0	252.4
pds80		62.3		132.5		107.2			154.1	164.4	196.6	198.0	260.1	259.5
pds90				108.5		105.1			166.3	169.4	195.2	195.4	267.9	265.7

2003) routines to improve our sparse matrix application based on hypermatrices. A fixed partitioning of the matrix has been done to be able to test the impact of each overhead reduction technique used. We present results obtained with and without bit vectors for two data submatrix sizes: 8×8 and 4×32. For the latter we also introduce the usage of windows.

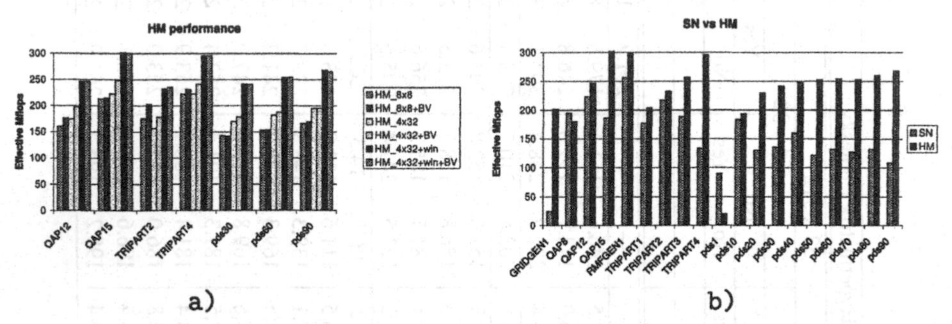

Figure 7. **a)** HM performance for several input matrices. **b)** SN vs HM performance.

Figure 7a summarizes these results. The usage of windows clearly improves the performance of our sparse hypermatrix Cholesky algorithm. We observe that the usage of bit vectors can improve performance slightly when windows are not used. When windows are used, however, bit vectors are not effective at all. Figure 7b compares the best result obtained with each algorithm for the whole set of test matrices. We have included matrix pds1 to show that for small matrices the hypermatrix approach is usually very inefficient. This is due to the large overhead introduced by blocks which have plenty of zeros. For large matrices however, blocks are quite dense and the overhead is much lower. Performance of HM Cholesky is then much better than that of the supernodal algorithm. This is due to the better usage of the memory hierarchy: locality is properly exploited with the two dimensional partitioning of the matrix which is done in a recursive way using the HM structure.

Finally, figure 8 shows performance of each algorithm on several matrix families. Note that, contrary to the supernodal algorithm behavior, the hypermatrix Cholesky factorization improves its performance as the problem size gets larger.

4. Conclusions

A two dimensional partitioning of the matrix is necessary for large sparse matrices. The overhead introduced by storing zeros within dense data blocks can be reduced by keeping information about a dense subset (window) within each data submatrix. Although some overhead still remains, the performance of our sparse hypermatrix Cholesky is up to an order of magnitude better than that of a supernodal block Cholesky which tries to use the cache memory prop-

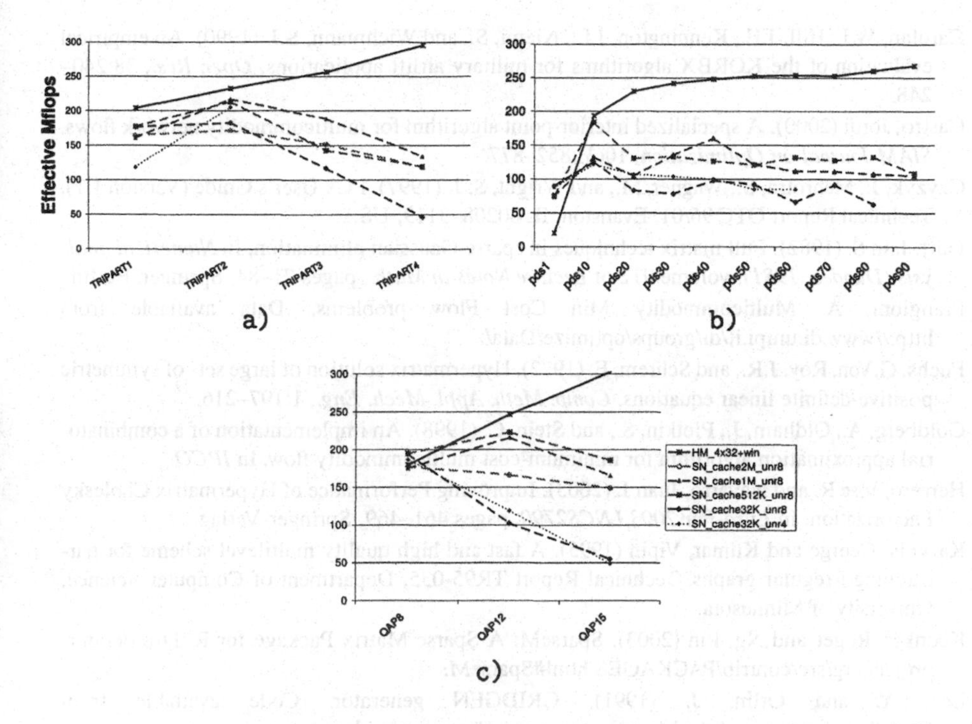

Figure 8. SN vs HM Cholesky for 3 matrix families: **a)** Tripart; **b)** PDS; **c)** QAP.

erly by splitting supernodes into panels. Using windows and SML routines our HM Cholesky often gets over half of the processor's peak performance for medium and large size matrices factored in-core.

Notes

1. However, if we look at the elements within those matrices we can see that the product $A \times B^t$ will produce a null update on C. In this case, the usage of bit vectors would be useful and could avoid this operation.

References

Ashcraft, C. and Grimes, R. G. (1989). The influence of relaxed supernode partitions on the multifrontal method. *ACM Trans. Math. Software*, 15:291–309.

Ast, M., Barrado, C., Cela, J.M., Fischer, R., Laborda, O., Manz, H., and Schulz, U. (2000). Sparse matrix structure for dynamic parallelisation efficiency. In *Euro-Par 2000,LNCS1900*, pages 519–526.

Ast, M., Fischer, R., Manz, H., and Schulz, U. (1997). PERMAS: User's reference manual, INTES publication no. 450, rev.d.

Badics, Tamas (1991). RMFGEN generator. Code available from ftp://dimacs.rutgers.edu/pub/netflow/generators/network/genrmf.

Carolan, W.J., Hill, J.E., Kennington, J.L., Niemi, S., and Wichmann, S.J. (1990). An empirical evaluation of the KORBX algorithms for military airlift applications. *Oper. Res.*, 38:240–248.

Castro, Jordi (2000). A specialized interior-point algorithm for multicommodity network flows. *SIAM Journal on Optimization*, 10(3):852–877.

Czyzyk, J., Mehrotra, S., Wagner, M., and Wright, S. J. (1997). PCx User's Guide (Version 1.1). Technical Report OTC 96/01, Evanston, IL 60208–3119, USA.

Duff, Iain S. (1982). Full matrix techniques in sparse Gaussian elimination. In *Numerical analysis (Dundee, 1981)*, volume 912 of *Lecture Notes in Math.*, pages 71–84. Springer, Berlin.

Frangioni, A. Multicommodity Min Cost Flow problems. Data available from http://www.di.unipi.it/di/groups/optimize/Data/.

Fuchs, G.Von, Roy, J.R., and Schrem, E. (1972). Hypermatrix solution of large sets of symmetric positive-definite linear equations. *Comp. Meth. Appl. Mech. Eng.*, 1:197–216.

Goldberg, A., Oldham, J., Plotkin, S., and Stein, C. (1998). An implementation of a combinatorial approximation algorithm for minimum-cost multicommodity flow. In *IPCO*.

Herrero, Jose R. and Navarro, Juan J. (2003). Improving Performance of Hypermatrix Cholesky Factorization. In *Euro-Par 2003,LNCS2790*, pages 461–469. Springer-Verlag.

Karypis, George and Kumar, Vipin (1995). A fast and high quality multilevel scheme for partitioning irregular graphs. Technical Report TR95-035, Department of Computer Science, University of Minnesota.

Koenker, Roger and Ng, Pin (2003). SparseM: A Sparse Matrix Package for R. http://cran.r-project.org/src/contrib/PACKAGES.html#SparseM.

Lee, Y. and Orlin, J. (1991). GRIDGEN generator. Code available from ftp://dimacs.rutgers.edu/pub/netflow/generators/network/gridgen.

Liu, J. H. W. (1990). The role of elimination trees in sparse factorization. *SIAM Journal on Matrix Analysis and Applications*, 11(1):134–172.

Liu, J. W., Ng, E. G., and Peyton, B. W. (1993). On finding supernodes for sparse matrix computations. *SIAM J. Matrix Anal. Appl.*, 14(1):242–252.

NetLib. Linear programming problems. http://www.netlib.org/lp/.

Ng, Esmond G. and Peyton, Barry W. (1993). Block sparse Cholesky algorithms on advanced uniprocessor computers. *SIAM J. Sci. Comput.*, 14(5):1034–1056.

Noor, A. and Voigt, S. (1975). Hypermatrix scheme for the STAR–100 computer. *cas*, 5:287–296.

Rothberg, Edward (1996). Performance of panel and block approaches to sparse cholesky factorization on the ipsc/860 and paragon multicomputers. *SIAM J. Sci. Comput.*, 17(3):699–713.

Rothberg, Edward and Gupta, Anoop (1991). Efficient sparse matrix factorization on high-performance workstations: Exploiting the memory hierarchy. *ACM Trans. Math. Soft.*, 17(3):313–334.

Zhang, Y. (1996). Solving large–scale linear programs by interior–point methods under the MATLAB environment. Technical Report 96–01, Baltimore, MD 21228–5398, USA.

IV

COMPUTATIONAL APPLICATIONS

PARALLEL IMAGE ANALYSIS OF MORPHOLOGICAL YEAST CELLS*

Laurent Manyri †
LAAS-CNRS, 7 Avenue du Colonel Roche, 31077 Toulouse, FRANCE

Andrei Doncescu
LAAS-CNRS, 7 Avenue du Colonel Roche, 31077 Toulouse, FRANCE

Laurence T. Yang
Department of Computer Science St. Francis Xavier University, Antigonish, NS, B2G 2W5, Canada

Jacky Desachy
GRIMAAG University of French West Indies, 97157 Pointe-ą-Pitre, FRANCE

Abstract Fermentation is a critical process for the production simpler substances from organic molecules or could be used to obtain the ethanol by the anaerobic breaking down of sugar. In this work we present different image analysis steps in order to characterize the cell morphology during the biotechnology process. The cell morphology is an important element of the stress which could disturb the production of the biomass or a metabolite. For this purpose we develop a Java software dedicated to an automatic analysis. The software allows us to have information about the growth cells, morphometric analysis (volume/surface) and morphology (budding cells). We want to have the information in biological real time to be able to modify the control parameters, therefore the image is cut into small slices which are analyzed separately by a parallel algorithm.

Keywords: Image Processing, JavaMPI, cells, snakes

* lmanyri@laas.fr
† Grant of Regional Council of Martinique

1. Introduction

Recently, the american researchers showed the human yeast carry out synthetically the otherwise different tasks of producing the human biosynthesis protein. The recent results allow us that a 'humanised yeast' could simplify drug manufacture by the introduction of human gene inside yeast chromosome. Since yeast grow faster and need less tending than mammalian cells, could be a possible solution to produce proteins cheaper and easier. One of the most famous example is the production of the human insulin for diabetics, which does not need added sugars, is brewed from yeast in this way. Moreover, it is difficult to produce these proteins to a commercial scale due to the incapacity of the biologist to keep the cell on a specific pathway therefore the necessity to develop control adaptive tools. Today, the pace of progress in fermentation is fast and furious, particularly since the advent of genetic engineering and the recent advances in computer sciences and process control. The high cost associated with many fermentation processes makes optimization of bioreactor performance trough command control very desirable. Clearly, control of fermentation is recognized as a vital component in the operation and successful production of many industries.

The last results indicated that the improvability and the control of the bioethanol production by alcoholic fermentation needs information about cell morphology.

The bioethanol is obtained by a chemical reaction :

$$C_6H_{12}O_6 \longrightarrow 2CH_3CH_2OH + 2CO_2$$

The operator of this reaction is a yeast, *Saccharomyces Cerevisiae* . The speed of this reaction and the production of ethanol depend of the physiological state of the cells (quality) and may be studied by morphometry (number of cells, surface and volume) and by morphology (classes). The yeasts develop by budding. Four classes of cells appear in the process : the single alive, the single dead, the budding cell (a mother and a child) and the aggregate of cells. This analysis is non based on a study of viability but we are interested in sizing and classifying the cells.
For this characterization, we develop different tools for microscopy image analysis in order to create an optical sensor which can be

- *off line in the case of separated treatment*
- *in situ for direct information on the process*

The aim of this study is to try to give the evolution of a cell during a process. Different kinetic models describe the steps of a alcoholic fermentation; this work have to approximate the steps of :

- adaptation : there's a change of the size of the cells
- exponential growth where the number of cells increase : increasing of the number of budding cells
- limit growth
- decline of the cells

2. Parallel Image Processing

In this paper we describe a new method of edge extraction of the yeast cells by active contours. To differentiate the budding cells of the single cells we have used the curvature analysis. The last algorithm used allowing to approximate each cell (budding or not) by an ellipse using a method based on least square method. All these steps are englobed in an unique algorithm which is applied in parallel for each cell or conglomerate of cells.

Image acquisition

One of the difficulty of the Cells Image Analysis is the modification of the size by take out of a sample. Two methods has been developed to solve this problem:

1 to examine under the microscope the cells

2 in-situ microscope

In this paper we present the first one. Therefore the growth of the cells during the fermentation is analysed by a microscope coupled to a camera which snaps several images. The microscope used is Olympus and the camera is a Nikon. A specific method of optical microscopy based on 'black background' allows to have only the contour of the cells. This method has been selected to reduce the influence of the background into the algorithm. The images obtained contain only the cell's edges with certain fuzziness.

Segmentation for parallel treatments

One of the control parameter for the 'health cells' is the modification of the ellipsoidal form which seems to be a stress indicator. If the computing time of this task is important the cell's culture risks to stop the proposed evolution. Due to the image size and time computation, the image is divided in small area : each area corresponds to one cell. This decomposition allows to use different system of processing. The final scheme chosen is a parallel implementation based on master/slave scheme. The master will segment the image and send a small image containing one cell to a slave. The computation of the slave is to

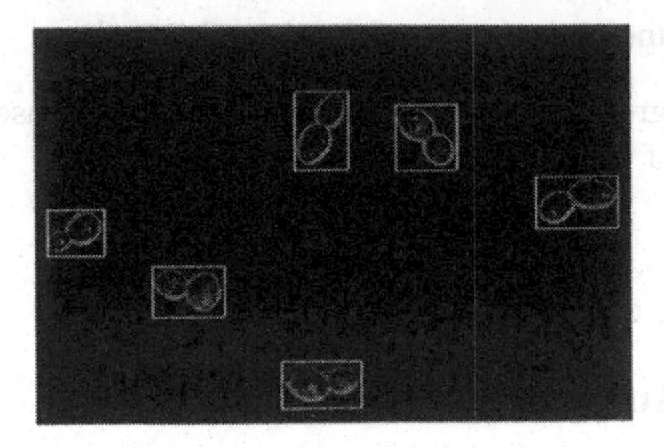

Figure 1. Separation of the cells

approximate the cell by a model and the master will recover the correspondent data for each cell in order to produce the stats. A simple threshold and separates the cells as it's visible on Figure 1.

JAVA MPI Interface

The treatment is divided by mean of the segmentation that identify region of interest (ROI). For these regions some subroutines have to estimate the contour of the cells by ellipses defined in section IV. As these subroutines used JAVA, we would like to use an approach to parallel programming in Java, defined as the most common object-oriented language.
For parallel processing, mpiJava (Carpenter, 1997) is an object-oriented Java interface to the standard Message Passing Interface (MPI, 1995); this interface was developed as part of the HPJava project. Indeed we need a good message passing API and Java presents various package for communication used in this interface that's why it's a well-designed, functional and efficient Java interface to MPI (Kanevsky, 1998).
Our parallel implementation is based on a master/slave scheme guided by mpi-Java.

Image segmentation by Snakes

One of the difficulty of the segmentation of this type of images is the diffraction on the cell membrane producing a fuzzy edge and of course the low level of luminance. In order to have a better approximation of the contour, we have ameliorated the convergence of the parametric snake, based on active contours proposed by Kass *et al.* (Kass, 1988). The snakes are used in computer vision and image analysis to detect and locate objects, and to describe their shape (Cohen, 1991).
A snake is a curved $x(s) = [x(s), y(s)]$, $s \in [0, 1]$, that moves through the

spatial domain of an image to minimize the functional

$$E = \int_0^1 (\alpha \|x'(s)\|^2 + \beta \|x''(s)\|^2) + E_{ext}(x(s))ds \tag{1}$$

- α and β are parameters that control the snake's tension and rigidity
- $x'(s)$ is the first derivative of $x(s)$
- $x''(s)$ is the second derivative of $x(s)$
- E_{ext} is the external energy function derived from the image, it rule is to move the snake toward the important values of the edges; for example, the gray value of the image, the negative first derivative may be used.

In our implementation, the external force is the gradient vector flow (Xu, 1997) (Xu, 1998). This field is computed as a spatial diffusion of the gradient of an edge map derived from the image. It's powerful than traditional snakes because they cannot move toward objects that are too far away and that snakes cannot move into boundary concavities or indentations.
How the cells are separated, each slave have to force the snake to lock on one contour. Different external forces have been used and to conclude, the gradient vector flow is the appropriate external force because of his fast convergence to concavities in the case of budding cell. This is the main condition to make the difference between the single cell and the budding cell.
At the end on the process a contour corresponds to a cell that is described in the map by (x_n, y_n) with $n \in [0, N]$.

3. Analysis

If the extraction of the cell has been done we need to find a robust algorithm allowing to measure the size of the single cell or the morphometric parameters of the budding cell. This involves that we are able to discriminate the mother cell and the daughter cell. The originality of this work is based on the curvature of the contour.
A single cell has a regular curvature; it should be approximate by a circle. In the other hand, a bud has an invariant curvature. For the curvature, we compute the radius as follow

$$R = \frac{\sqrt{d_{x_n}^2 + d_{y_n}^2}^{\frac{3}{2}}}{d_{x_n} d_{y_n}^2 - d_{y_n} d_{x_n}^2}.$$

But the change of the curvature may be appreciated by the sign of

$$d_{x_n} d_{y_n}^2 - d_{y_n} d_{x_n}^2 \quad \text{with}$$

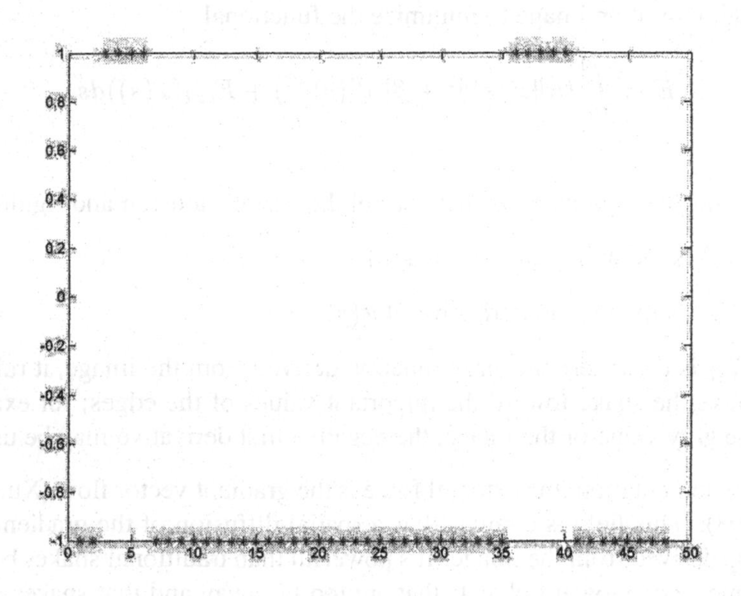

Figure 2. Curvature radius computation

- d_{x_n} and d_{y_n} the first derivatives of (x, y)
- $d^2_{x_n}$ and $d^2_{y_n}$ the second derivatives of (x, y)

To compute the first and the second derivative, the choice of the step allows or not to find the change of curvature. More it's small more there will be changes while more it's large less a change in the curvature can be detected. In this case of images, a step of 10 has been chosen.
We can evaluate the results of this computation on the Figure 2.

The X-axis correspond to the pixels and the Y-axis is the value (-1 or 1) of the radius. For a simple cell, there is no change of curvature but for this contour two peaks appear. When the curvature value is equal to -1 it's correspond to a cell (mother or child). Indeed the pixels corresponding to the first at the second peak is a cell; moreover the pixels between the second and the first peak belong to the second cell.

4. Cell Modeling

The biological knowledge allows us to modeling the cell shape by an ellipse. After the analysis of the contours there is no information a priori to distinguish a single cell of a budding one. This comparison is possible using the size of these cells; that's why this contours are approximated by models. A cell is considering an ellipse defined by five parameters : a center (x_0, y_0), a minor axis a, a major axis b and an orientation θ. It can be described by several equations :

- explicit $\frac{((x-x_0)+E(y-y_0))^2}{a^2(1+E^2)} + \frac{((y-y_0)+E(x-x_0))^2}{b^2(1+E^2)} - 1 = 0$
- implicit $F(a,x) = ax^2 + bxy + cy^2 + dx + ey + f = ax$ with $a = (a,b,c,d,e,f)$ and $x = (x^2, xy, y^2, x, y, 1)^T$
- parametric $< \overline{x} - \overline{v} > E < \overline{x} - \overline{v} >^T = r^2$

Different global and local methods have been evaluated, a global estimation of their efficiency is presented in Table .

Method	Time (sec)	Precision
Hough Transform (Hough)	360	Lower
Fuzzy c-shells (Dave, 1992)	2	Lower
Differential Evolutionary (Storn, 1995)	188	Fine
Stochastic gradient	0.01	Lower
Direct Least Square (Fit, 1999)	0.01	Very fine

Table 1. Precision and time computation for estimation

This table presents five methods that have been experimented in order to approximate a simple contour and the more efficient is Direct Least Square; it's described in the next section. The precision of one method is difficult to be appreciated; it represents the best value of the minimum of the sum of squared algebraic distance. Only the Differential Evolutionary Algorithm, a class of stochastic search and optimization methods including Genetic Algorithm, and the method Direct Least Square are efficient in time for this problem.

Fitting Ellipse

This is an efficient method for fitting ellipses from scattered data. Other algorithms either fitted general conics or were computationally expensive. By minimizing the sum of squared algebraic distance

$$D_A(a) = \sum_{i=1}^{N} F(x_i)^2$$

of the curve F to the points using the implicit equation of the conic :

$$F(a,x) = ax^2 + bxy + cy^2 + dx + ey + f = ax$$

To avoid the solution $a = 0$ some constraints are applied to the parameters :

- linear : $c.a = 1$
- quadratic $a^T Ca = 1$ where C is a 6x6 constraint matrix.

As a quadratic constraint is applied, the minimization can be solved by : $D^T Da = \lambda$ where $D = [x_1 ... x_N]^T$ and λ is the Lagrange multiplier. Moreover, as the conic is an ellipse, we impose $b^2 - 4ac > 0$.
Then $a^T Ca = 1$ as

$$a^T \begin{bmatrix} 0 & 0 & 2 & 0 & 0 & 0 \\ 0 & -1 & 0 & 0 & 0 & 0 \\ 2 & 0 & 0 & 0 & 0 & 0 \\ 0 & 0 & 0 & 0 & 0 & 0 \\ 0 & 0 & 0 & 0 & 0 & 0 \\ 0 & 0 & 0 & 0 & 0 & 0 \end{bmatrix} a = 1$$

Then the minimization is about $E = \|Da\|^2$ with the constraint $a^T Ca = 1$. Introducing Lagrange multiplier the system becomes

$$2D^T Da - 2\lambda Ca = 0 \quad (2)$$
$$a^T Ca = 1 \quad (3)$$

Then

$$Sa = \lambda Ca \quad (4)$$
$$a^T Ca = 1 \quad (5)$$

where $S = D^T D$. The system is solved by considering the generalized eigenvectors of (4).
If (λ_i, u_i) solves (4) then $(\lambda_i, \mu u_i)$ for any μ from (5), we can find μ_i as $\mu_i^2 u_i^T Cu_i = 1$ giving

$$\mu_i = \sqrt{\frac{1}{u_i^T Cu_i}} = \sqrt{\frac{1}{u_i^T Su_i}}$$

5. Results and discussion

The image's segmentation, the analyze of the contours and the approximation of elliptic model allow to count the number of singles and budding cells. Note that after the analyze of contours, the cell may correspond to a budding cell i.e. a mother and a child or it may correspond to two singles. Indeed, by modeling and having the parameters, the two major axises are compared. Consider $a_{max} > a_{min}$ the two axes then :

- if $a_{max}/a_{min} > 1/3$ there are two cells
- if $a_{max}/a_{min} \leq 1/3$ it's a budding cell i.e. a mother and a child

The cells (single and budding cells) have been counted in the Figure 3. As describe in the first section, the traditional schemes show an exponential growth

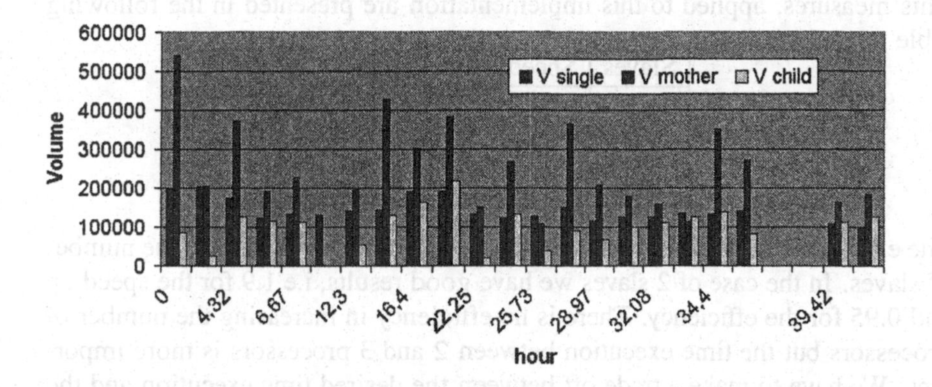

Figure 3. Percentage of single and budding cells

: an increasing of the number of buds. These methods have been implemented on a Local Area Network using MPI, interfaced by JavaMPI. Our strategy uses laboratory clones (Pentium III 450 Mhz, 128 Mo).

Slaves	1 image	1500 images
1	40sec	60000sec
2	21sec	31500sec
3	16sec	24000sec

Table 2. Time computation for slaves

This table presents time and estimated time execution for different slaves number. We can estimate the time execution for N images because it's a linear function. For off line treatments, the time is not important but in the case of on line estimation, we'll need more powerful clusters. Indeed if we implement these methods on a Pentium M (1.3GHz, 256Mo) the execution time is 10 sec for an image; it will be implemented in future works. Our goal for the fermentation process for a real time sensor is to characterize the cells physiological states by treating 100 images in 3 minutes.
If we consider p the number of processors or clones, T_s the sequential time execution, $T(p)$ the time execution for p processors, $A(p)$ the speed up or the acceleration and the efficiency $E(p)$ we have :

$$A(p) = \frac{T_s}{T(p)} \text{ and } E(p) = \frac{A(p)}{p}$$

This measures, applied to this implementation are presented in the following table

Slaves	Speed up	Efficiency
2	1.9	0.95
3	2.5	0.83

Table 3. Efficiency of parallelism

The efficiency has to be near to 1 and the speed up has to be near to the number of slaves. In the case of 2 slaves we have good results, i.e 1.9 for the speed up and 0.95 for the efficiency. There is no efficiency in increasing the number of processors but the time execution between 2 and 3 processors is more important. We have to make a trade off between the desired time execution and the efficiency of the implementation.

6. Conclusion

The databases concerning the yeast increase each year with new results about the discovered of new proteins and new behavior due some expressed genes. The mathematical modeling of the eukaryote cells therefore the yeast is always an open question which excites the scientific community. One of the parameters which could be used by the biologist to understand the yeast evolution is the morphometry of it. These parameters could be introduced in a structural model allowing to analyze the cell cycle.

We present in this paper a robust algorithm for morphological yeast analysis allowing to have the information about the budding and stress modification in biological real time. This is due to our parallel software which has been based on a master/slave scheme guided by JavaMPI. In future works

7. Acknowledgments

The authors would like to thank Fermentation Team of INSA Toulouse Laboratory of Biotechnology for the experimental data used in this paper.

References

R. N. Dave and K. Bhaswan, *"Adaptive fuzzy c-shells clustering and detection of ellipses"*, IEEE Transactions on Neural Networks 3, pp. 643-662, September 1992.

Andrew W. Fitzgibbon, Maurizio Pilu, and Robert B. Fisher, *"Direct least-squares fitting of ellipses"*, IEEE Transactions on Pattern Analysis and Machine Intelligence, 21(5), 476–480, May 1999

P. V. C. Hough, *"Method and means for recognizing complex patterns."*, U. S. Patent 3,069,654, 1962.

Michael Kass, Andrew Witkin, and Demetri Terzopoulos, *"Snakes: Active contour models"*, International Journal of Computer Vision, pages 321-331, 1988

Storn R. and Price K., *"Differential Evolution - A simple and efficient adaptative scheme for global optimization over continuous spaces"*, Technical Report TR-95-012,ICSI 1995.

C. Xu and J.L. Prince, *"Gradient Vector Flow: A New External Force for Snakes"*, Proc. IEEE Conf. on Computer Vision Pattern Recognition (CVPR), Los Alamitos: Comp. Soc. Press, pp. 66-71, June 1997.

C. Xu and J. L. Pince, *"Generalized gradient vector flow external forces for active contours"*, Signal Processing , 71 (1998) 131Ű139.

L. D. Cohen, *" On active Contour Models and Balloons"*,CVGIP: Image Understanding, 53(2):211-21, March 1991

Bryan Carpenter, Yuh-Jye Chang, Geoffrey Fox, and Xiaoming Li, *" Java as a language for scientific parallel programming"*', In 10th International Workshop on Languages and Compilers for Parallel Computing, volume 1366 of Lecture Notes in Computer Science, pages 340-354, 1997.

Arkady Kanevsky, Anthony Skjellum, Anna Rounbehler, *" MPI/RT - An Emerging Standard for High-Performance Real-Time Systems"*, HICSS (3) 1998: 157-166

Message Passing Interface Forum, , *" MPI: A Message-Passing Interface Standard"*, University of Tenessee, Knoxville, TN, June 1995. http://www.mcs.anl.gov/mpi.

AN HYBRID APPROACH TO REAL COMPLEX SYSTEM OPTIMIZATION

Application to satellite constellation design

Enguerran Grandchamp
University of the French West Indies
enguerran.grandchamp@univ-ag.fr

Abstract This paper presents some considerations about hybrid optimization algorithms useful to optimize real complex system. The given indication could be used by readers to conceive hybrid algorithms. These considerations have been deducted from a concrete application case: the satellites constellations design problem. But each of the advanced techniques proposed in this paper are considered in a more general way to solve other problems.
This problem is used to illustrate the techniques along the paper because it is grouping many characteristics (difficulties) of contemporary real complex systems: the size and the characteristics of the search space engendered by a combinatorial problem; The irregularity of the criterions; The mathematical and physical heterogeneity of parameters forbids the use of classical algorithms; The evaluation of a solution, which uses a time consuming simulation; A need of accurate values. More details are available in previous papers (10, 13, 11, 12). For these reasons, we could learn a lot from this experiment in order to detach hybrid techniques usable for problems having close characteristics. This paper presents the historic leading to the current algorithm, the modeling of the complex system and the sophisticated algorithm proposed to optimize it.
Application cases and ways to built significant tests of hybrid algorithm are also given.

Keywords: Hybrid optimization, metaheuristics, modeling, simulation, constellation design

1. Introduction

This paper explores a hybrid meta-heuristic approach to solve complex systems optimization problems. These methods are dramatically changing our ability to solve problems of practical significance. Faced with the challenge of solving hard optimization problems that abound in the real world, classical methods often encounter great difficulty even when equipped with a theoretical guarantee of finding an optimal solution. So we use such algorithm as classi-

cal resolution techniques found their limits, typically for large combinatorial exploration space.
This paper is more centered on optimization than on space system design. The main goal is to underline and illustrate the power of hybrid optimization and the way to introduce expert knowledge to obtain a more efficient search.
Nowadays, complex systems are defined by a large number of parameters and operational constraints induce the necessity of accurate values.
These two goals are on opposition and exploring the whole searching space with accuracy is not feasible in most cases: using techniques to explore a lot of areas from the searching space doesn't allow to produce accurate values, using a local optimization to refine the values doesn't allow to drive a global optimization.
In such a case, hybrid optimization gives an elegant solution. We have to intelligently manage the two approaches to refine values in the neighborhood of good local optima and to privilege the exploration in less important areas. The orchestration is the key of success and we have to exploit the best of each technique at the appropriate instant.
The paper is organized as follows.
Section 2 introduces the problem to solve in its general context. It explains why an optimization process is required with technical and economical reasons. We also precise the limits of the studied system, the evaluation process of a solution and the research space.
Section 3 presents a frequently adopted approach by the Constellation Designer Community. We will analyze its limits to find a new way to solve the problem.
Section 4 presents the new approach foundations. It details the main ideas used to design the algorithm (physical signification of the parameters, guiding the search, accurate optimization).
Section 5 integrates all these remarks to present the whole algorithm. All stages are detailed in this section: the knowledge database, both high and low level stages. A special attention is given to the orchestration level.
Section 6 presents the tests and applications. We will first validate each stage separately before to make a global validation. Both telecommunication and navigation application fields are considered.
Section 7 deals with operational use of the techniques. It presents the developed software and the way to use it. We also give indications to speed up the algorithm.
Section 8 closes the paper with some conclusions and perspectives.

2. Satellite constellation optimization background

The presentation of the problem in its economical (decreasing the costs) and technical context (increasing the performances) is necessary to understand

the challenge of the optimization process and the place of design phase in the whole spatial system.
Halfway between optimization and astronautic the satellites constellations design problems deals with many goals. To find the required number of satellites and to correctly set their position, such are the technical challenges of this study. To minimize the cost and reduce the time such are the economical challenges the space industries are confronted in a daily manner (37).
In this section, we shortly detail the whole space system and the limits of the studied system. Then we detail its parameters to show the complexity of the problem: numerous parameters with different physical and mathematical nature.

Space systems: towards a sequential optimization

Satellite constellations are currently used in many fields for civil and military applications: earth observation, telecommunication, navigation. The implementation of a constellation is a multi step process: mission analysis, design of the constellation, deployment of the satellites, satellite checking, maintenance, replacement and termination. Each of these steps must be optimized to decrease the global cost and increase the global performance.
There are two approaches to conceive such a system: a global approach which integrates every step in a multi-objective optimization; a local approach which optimize each step separately.
The first way to solve the problem is currently not practicable for many reasons: firstly the nature of the problems to solve in each step is too different to regroup them in a general algorithm. As an example the deployment step looks like a Constraint Satisfaction Problem (CSP), where some constraints are satellite launcher availability, payload, cost and accessible altitude. The evaluation function integrates cost and time parameters. The design step is a combinatorial problem. The aim of this step is to minimize the number of satellite (in order to reduce the cost) and to maximize the performances. Secondly the number of parameters to set is too large with integer and real variables. For these reasons, the second approach is preferred (36) and each step is optimized in a different manner. In fact, a step optimization is not totally uncorrelated from the other one.
As if we optimize only one step, some considerations about the previous or next step could influence the current optimization process (as constraints or guide).

Limits of the system

The system we want to optimize is limited to the constellation. We have to set the parameters for each satellite that compose the constellation and also to

determine the number of satellites to use.
A satellite is defined by six orbital parameters $(a, e, i, \omega, \Omega, M)$ (precise definition is given in 26), having physical signification and unit.

- a: semi axis of the orbits ellipse (from 400 km to 36 000 km)
- e: eccentricity of the orbits ellipse (0 to 1)
- i: inclination of the orbit plane (0 to 360 degrees)
- ω: argument of the perigee (0 to 360 degrees)
- Ω: longitude of the ascending node (0 to 360 degrees)
- M: mean anomaly (0 to 360 degrees)

A constellation is defined by N satellites.
Without additional constraints there is no relation between satellites and the system is defined by 6N independent parameters.
Moreover, the number of satellites (N) is also unknown and the problem has a variable number of parameters.

Solution evaluation

The efficiency of a constellation should be guaranty during its revolution period. The duration of the period depends on the orbital parameters of each satellite and the evaluation is based on an accurate simulation process.
Usually this simulation is time consuming and depends on many operational parameters. In fact, ground sampling could change from 10 km to 100 meters and time sampling from 5 minutes to few seconds. The performances could be awaited for the whole earth surface or just over certain areas (north hemisphere, Europe, ...
For each time-space sample we evaluate a local performance value. This value could be the number of satellites visible from the sample (telecommunication application), the positioning precision induced by the local configuration of the visible satellites (navigation application) or any other technical value.
From this amount of local evaluations, we compute few values (mean, maximum, minimum) to estimate the constellation performance.
As an example, we used the precision function to optimize a navigation constellation (such as GPS or GNSS) with the simulation parameters set as follows

- The time step is set to 1 minute because it is the maximum acceptable gap between two position values for operational conditions.
- The ground sample step is set to 10 km around the whole earth (satellite localization must be available everywhere). This value has been chosen

because it is sufficient for an accurate computation and acceptable for time consumption. But with such value a second validation is necessary with a ground sample step at least of 500 m after constellation optimization.

- The local criterion is a triangulation compute in order to evaluate the local precision. From satellites positions an iterative algorithm is run to minimize the positioning error (least square method).

- At the end of the simulation process, minimum precision is returned in order to evaluate if the constellation is acceptable, mean precision could also be returned.

The time to compute is about 30 seconds on a Pentium IV 2Ghz. We could already note that there is a lack of information returned by an evaluation compared to the time spends during the simulation.

Research space

The research space is generated by the 6N parameters. N is an integer and the 6N parameters are real variables.
Theoretically the searching space is infinite. In some optimization contexts we conserve compact intervals. This is the case when defining a parameter as a combination of others. For example, if we want an orbit with a one day period, there is a relation between a an i (26).
In most application cases, a sampling of each interval is made. The problem is converted to a combinatorial problem with a very large searching space. In fact, the sampling step could be of 100 meters for a, 0.01 for eccentricity and 0.1 degrees for each other range.
In such a context the size of the space has about 6.10^{21} elements $(356000 * 100 * 3600^4)$ for each satellite.
For a constellation with N satellites the searching space has $(6.10^{21})^N$ elements.
For a constellation with an unknown number of satellites included between 1 and N the searching space has about $(6.10^{21})^{\frac{N(N+1)}{2}}((6.10^{21})^1 + (6.10^{21})^2 + ... + (6.10^{21})^N)$. N could be equal to 100.
With such a searching space, an exhaustive search is impossible.

3. The first approach: an instructive failure

Principe

In front of a complex problem with many parameters, a large searching space, an expansive evaluation and without rules to predict the impact of a modification to the value of the solution, the first idea is to use an algorithm

which is reputed to treat any kind of optimization problem with a statistical guaranty to reach the best solution and without any consideration about the nature of the problem.
As if there is no universal algorithm there is some algorithm quiet easy to parameter for any kind of problem. Genetic algorithms are part of them. The main advantage is that the research process is a standalone process using basic operators like selection, crossover or mutation.
With such an algorithm, the resolution is theoretically easy. But when considering the wide search space, the population has to be very large (several thousand) in order to leads to a significant global optimization.
When adding the time consuming criterion, this way looks unfeasible and the first idea is to reduce the complexity of the search space.

Reducing complexity: exploration of regular or symmetric constellations

Usually, engineers reduce the complexity of the search space by linking parameters. We artificially reduce the number of parameters to optimize. For example, fixing a common altitude, inclination and eccentricity reduce the parameters from 6N to 3(N+1).
Such approach reduces the kind of solution the algorithm could reach. In this way, only symmetrical solutions are explored (same kind of orbit, ...).
This is the case of Walker constellations (57) which are composed of satellites having the same altitude and inclination with a uniform distribution of the orbits around the earth and a uniform distribution of the satellites around the orbits.
With such constraints, the optimization algorithms are skewed and willingly designed for symmetric resolution.

An incomplete panel of solution

Such symmetric solutions have intrinsic properties. We could easily design a constellation that satisfied the required local performances but without any guaranty about the number of satellites (which is directly linked to the cost of the constellation).
Moreover some recent constellations (28,30) are composed of different kind of orbits (tundra, circular, heliosynchronous) and are not accessible from a symmetric point of view.
Designing a symmetric constellation as a starting point for the optimization could appear to be a good idea but the local optimum created by this constellation is so attractive that we must totally break the symmetry to access other kind of solution.
More details on classical approaches are available in the literature (31, 33, 35).

4. The new approach foundations

Born from the limits and drawbacks of the previous method and several preliminary studies, this approach tries to answer to all expectation in a more constructive way than the previous one.
Based on a better use of the simulation and on a simplification of the criterions, the algorithm is composed of several levels and uses different optimization techniques: it integrates a knowledge database on the orbits and a numerical optimization process both orchestrated by a metaheuristic algorithm.
The main idea is to bypass the drawbacks with a decomposition of the system in order to adapt the algorithm to the nature of the sub-problem.
When dealing with a wide, dark and compact search space, we have to find a way to introduce an "exploration map", a "spot light" and a "magnifying glass". Following subsections describe the way to materialize this representation.

Physical signification of the parameters: splitting the search space (the spot light)

For each complex system we advise to make a particular effort to precisely understand the influence of each parameter to the system behavior.
If we consider the physical signification of the parameters we could classify them according to their importance or influence to the solution.
A small variation on a parameter could dramatically change the solution evaluation or on contrary a coarsely variation could have a least effect.
In our application, the physical differences of the parameters previously underlined as being a drawback for a general optimization process without precise consideration will become an advantage in the new modeling.
In fact, for some parameters (a, i, Ω) we could easily analyze their impact on a single orbit. The information given is not extensible to predict the interaction with other orbits because the main problem becomes the synchronization of the satellites.
For example, the parameter a has a direct influence on the area visible from the satellite and the parameter i is linked to the accessible latitude.
So we could divide the parameters ranges into logical subset.
These considerations are at the origin of the expert knowledge introduced in the next section.
The advantage of such an approach is to reduce the complexity (combinatorial) without reducing the panel of solution.
So the better way we found is to organize the search in order to split the space into logical areas. If these areas are intelligently conceived, we could evaluate the ability of each one to resolve the problem and so dramatically reduce the exploration.

This will become an essential actor of the algorithm success and the way to bring light to the search.

An efficient progression: guiding the search to avoid blind optimization (the exploration map)

As if we intelligently organize the search space, there are a lot of areas to explore. We have to manage this exploration in order to guide or constraints it. We use information given by the previous study (parameters) to select the areas, information given by users to constraint the search in a selected way, information returned by previous search. The goal of this part of the algorithm is twice: driving the search among the different areas and coordinates the ratio between local and global search.
Advanced metaheuristics techniques will be used to ensure the efficiency of the progression.

An accurate optimization: local intensification of the search (the magnifying glass)

After the localization of the interesting area, we have to precisely set the value of each parameter. This part of the optimization process deals with fine local optimization.
Specifications are not the same for the algorithm to employ. It must be able to reach a local optimum very quickly. The techniques used are close to classical optimization.
A special reflection must be done in order to integrate the possibility of multiple local optima in the same area (in the sense of previous subsections).
In fact, additionally to previous drawbacks, the criterion used to evaluate a solution usually presents some unfriendly characteristics: many local optima, chaotic surface.
Combining with previous algorithm, we hope being able to explore a wide space and to precisely analyze only interesting one.

Learning from the simulation: converting evaluation time to optimization time.

As previously underlined, the evaluation criterion is time consuming and produces few values. Moreover the simulation doesn't return pertinent information about the good or bad properties of the satellites. These values don't allow differentiating satellites within the constellation in order to modify the orbit of less efficient one.
The idea is to use the simulation process to compute many other values for each satellite in order to evaluate their contribution to the efficiency of the con-

stellation.
This computation induces a fee to the evaluation function but the pertinence of the gain justifies the effort. Moreover the overload is not proportional to the number of values because some of them are computed for each time step instead of for each time-space sample.
For example, a value (per satellite) could be the duration of visibility from the interesting area (telecommunication or earth observation system) or the number of time a satellite is implied in a high precision positioning triangulation (navigation system).
These values will be used during the main algorithm presented later in the paper in order to guide the progression.

5. Modeling: Towards a multi stage optimization

When considering all previous remarks and advises a multistage hybrid algorithm comes naturally. In fact, there is a local and a global approach, an accurate optimization and a wide exploration need. Problems to solve are totally different and must be parsed in order to be treated.
Other hybrid approach are presented in the literature (38). We define three stages for the algorithm: a numerical stage dealing with accurate optimization in a local context; a heuristic stage introducing expert knowledge to organize the search space; a metaheuristic stage managing the progression of the search. Each stage is detailed in the next subsections.

The knowledge interconnection stage: the orbit database

The Orbit Data Base (ODB) is the tool that allows restricting and organizing the search space. It is the way to introduce expert knowledge by the definition of Orbit Classes. According to the characterization of an orbit (Section 0), each class defines a subset of values for each of the six orbital parameters. In fact, a class regroups orbits with common characteristics (altitude, periodicity, inclination, eccentricity, periodicity ...). That is to say, a class is combination of ranges (13, 19) for some orbital parameters, fixed values or relations for others. As an example, if we want to represent orbits which have a revolution period of one day (useful for a daily earth observation) we define a relation between parameters a, e and i. In such a case, the class is not composed of successive orbits.
Now we directly manage classes instead of satellites (a satellite is a sample of a class). This naturally organizes and split the search space into two hierarchical subspaces: subspace of classes and subspace of satellites within a class. The subspace covert by the set of classes is not necessarily the whole search space (search space engender by the accessible orbit defined by the six orbital parameters). Moreover, classes are not necessarily disjointed.

In fact, in standard classification (16), each sample has to be assigned to exactly one class. The ODB relaxes this requirement by allowing gradual memberships, offering the opportunity to deal with satellites that belong to more than one class at the same time.
This fundamental characteristic of the ODB is extended with fuzzy concepts (17, 18, 19) allowing a flexible frontier between classes.
In our application case, the introduction of fuzzy classes is necessary for many reasons.
First of all, many orbital classes haven't a well defined border. For example, we frequently talk about Low Earth Orbit (LEO) to represent orbits starting with an altitude of 400 km. We also talk about Medium Earth Orbit (MEO) to represent orbits near 12 000 km altitude. But what about an orbit with 6 000 km altitude? Is it a LEO or a MEO one? The transition between LEO and MEO is not so easy to express and the use of fuzzy borders allows a smooth transition between classes.
Secondly, this property appears very useful during the progression of the search because the definition of this intersection between classes allows a natural transition between them.
In fact, when changing a solution, it frequently appends to reach the border of a class. In such a case, we exchange the current class with one that contains the solution (intersection not empty).
Figure 1 illustrates these characteristics. The number and the nature of the

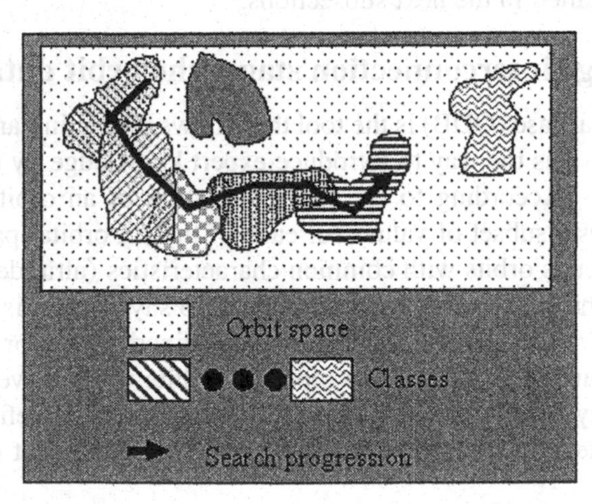

Figure 1. Search space decomposition

classes are determinant for the efficiency of the search and the characteristics of the returned solution.
The parameters that constraint the different stages (number of iterations, thresh-

olds, ...) will also depend on the classes definition.
In fact, as an example, if the ODB is composed with a lot of small classes, an effort will be made for the high level stage to choose the representatives classes and their proportion. On contrary, with less but larger classes, the low level stage will be used to refine the parameters of each satellite.
Now a satellite becomes a handler of a class (the satellite parameters take their value in the fuzzy range defined by the belonging class).
As a constellation is a combination of satellites, a configuration is defined as a combination of ODB class handlers. The fundamental difference between a constellation and a configuration is that one configuration contains a set of constellation (each constellation whose satellites verify the corresponding ODB class). We could compare the configuration space to the constellation space in the same way as in section 2.
Examples of orbits are given in Figures 2-a, 2-b, 2-c, 2-d. in order to illustrate the panel of accessible classes. When designing the ODB, we have to define a

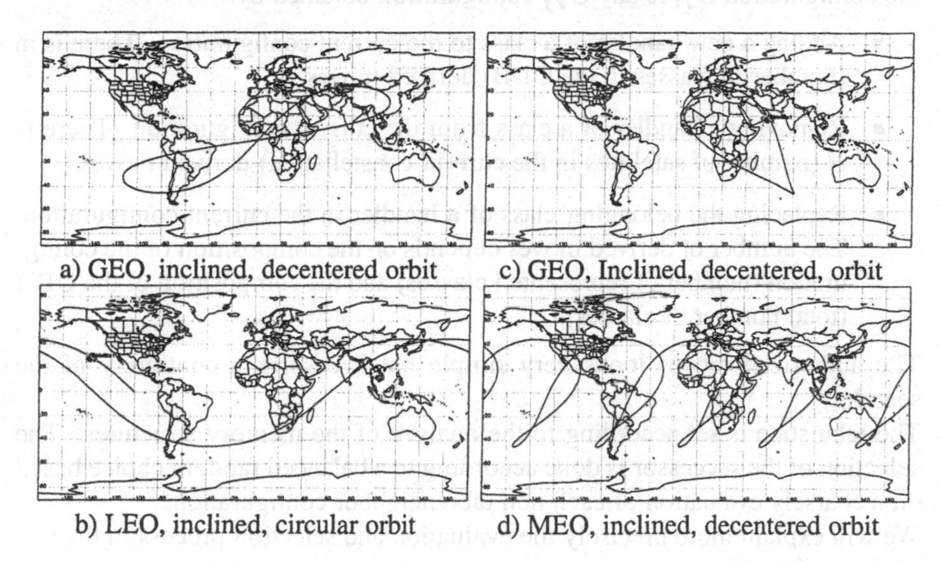

a) GEO, inclined, decentered orbit — c) GEO, Inclined, decentered, orbit

b) LEO, inclined, circular orbit — d) MEO, inclined, decentered orbit

Figure 2. Orbit sample

complete package to manipulate the classes. In a general way, it includes the definition of the elements (accessible orbits), an order to browse them and an operator to compare them. We also have to define a notion of distance between classes, orbits and configurations.
For more details readers are invited to refer to previous works (13).
We note Cl_k a fuzzy orbit class, m the number of classes, $Cl_{k\,k=1..m}$ the ODB and Cfk a configuration.

High level stage: the case of Tabu Search

Tabu foundation This level tries to answer the question "which configuration to explore and how to explore it?". The goal of this level is to manage the search.
The Tabu Search method (3,4,5) has been retained because it is a local search method avoiding heaviness of managing a population of solution and because it integrates the notion of history through memory structures.
This algorithm is very complete and enables learning from previous exploration to predict future progression.
The neighborhood used for the high level stage is composed of configurations. At an iteration of the algorithm, we have to choose the successor of the current confirmation between the configurations that composed its neighborhood.
As it is a local search, we define the neighborhood as follow. A neighbor of the configuration Cf_i is any Cf_j configuration obtained by

- Adding a new handler of a class to the current configuration. There is m (number of classes in the ODB) derived moves.
- Removing a handler of a class from the current configuration. There is N (number of satellites in the current constellation) derived moves.
- Replacing the belonging class of a handler in the current configuration. The number of derived moves depends on the composition of the configuration (number of represented classes) and the composition of the ODB (total number of classes).

The authorized moves looks very simple but it ensures a continuity of the search.
The tabu state is set according to the analyze of the memory structures. The selection of the successor is done according to a balanced random choice based on a coarsely evaluation of each non tabu neighbor configuration.
We will explain more precisely the evaluation and selection process in the rest of the section. *M*emory management
The inspiration of Tabu Search to design the high level stage leads us to introduce different kinds of memory structures.
The memory aspect of Tabu Search is essential to drive the search on good conditions.
Short term and long term memory are present in Tabu theory and accompanied by their own special strategies.
The effect of both types of memory may be viewed as modifying the neighborhood of the current solution.
Each memory structure is a kind of history filter which memorize interesting characteristics. We implemented all these structures to have a complete algorithm.

Tabu history analyzing is based on both short term memory (recency-based memory) and long term memory (Quality, Influence, Frequency).
Short term memory forbid to visit recently explored solutions. Long term memory allows to analyze the good or bad effect of certain choice.
The Quality aspect of the memory is used to qualify the contribution of transition to the improvement of a solution.
In our application case, if the addition of a Clk handler often improve the evaluation of the solution (increase performances and/or decrease the costs), the corresponding class will be notify with a high quality value.
The quality value could be affected to each elements of the system: a class, a configuration, a satellite, a transition.
For a configuration there are two steps in the attribution of a quality value. First choosing a configuration among neighbor means evaluating them before any exploration (if we don't want to make a random choice). Secondly, after the exploration of the configuration with the low level algorithm we could precisely evaluate it.
The post evaluation of the configuration is a kind of mean constellation value.
The evaluation of a class is a balanced mean of the configuration values where the class is represented.
The pre evaluation of a configuration is based on the evaluation of the classes represented in the configuration.
Then a satellite evaluation is directly linked to the simulation process presented in section 2. We compute separated values for each satellite in order to learn from the simulation to manage the memory.
More details about the expression of the different values are given in (13).
The frequency memory aspect gives information about the number of time we use a move (or a derived move). This indication is used for diversification or intensification process (developed in the continuation).
We could already note that the exploration history take a predominant part in the quality memory management.
The consequence is that the search is starting with a masked neighborhood (random exploration) and values become more and more precise as the search is going on.
High level decisions (exploring a configuration or another) are more and more pertinent and the search is expected to be more and more efficient.
To avoid the blind starting search, we advise to adopt a kind of learning phase in the neighborhood of the starting solution (configuration). This learning phase consist of exploring as much configuration as possible in order to give a first idea of the quality evaluation.

Low level stage: the case of Steepest Descent

The low level stage has been designed to quickly reach a local optimum in a restricted area. This area is limited by the current configuration (authorized values for a satellite within its belonging class).
The goal is not to explore every constellation within the current configuration but to rapidly reach a good solution in order to evaluate the potential of the configuration to solve the problem.
At the origin, this algorithm was a simple Steepest Descent. This algorithm has been chosen as if it could encounter slow convergence because of simplicity of implementation. But the criterion used for our applications is irregular and present many local optima.
So we propose an evolved Steepest Descent algorithm with a restarting strategy in order to bypass the local maxima.
The restarting strategy is a random modification of some orbital parameters. Parameters are selected according to their progression during the steepest descent phase. We select the parameters having changed the least during this phase.
We should note that this restarting strategy is a kind of diversification phase from the low level point of view but stays a local intensification phase from the previous high level stage.
This low level is not the main part of the algorithm because it has a precise and restricted role but its operational implementation is not necessarily easy.
Memory is also present at low level optimization. In fact, during the search it is possible to come back to a configuration already visited. In such a case, we have to restart the search from the last visited solution or from the best encountered solution.

Precisely tuning low and high level optimization parameters: the role of Strategic Oscillations for intensification and diversification ratio.

When analyzing the whole algorithm, we could note the multitude of parameters. The low level stage includes a threshold for the detection of a local optimum, a number of iterations, a progression step .... The high level stage includes a threshold on class values for the tabu state, a neighborhood size, a range for the number of usable satellites ...
Fixing a value for each parameter is not so easy for many reasons. Firstly, the thresholds and ranges of each parameter strongly depend on the use case. We calibrate the algorithm for each space system application (navigation, earth observation and telecommunication).
Secondly, the optimization needs change during the search. At the beginning, we have to explore several areas in order to locate the most interesting one be-

fore to deepen more accurately few configurations at the end of the search. Moreover, the parameters depend on the search progression itself. In case of a satisfying search, an intensification phase is preferred in order to exploit the local area. On contrary, if the search is not concluding, a diversification phase starts to reach other areas.
This part of the algorithm is the key of the success. It coordinates the different techniques and algorithm to apply the most adapted one.
We use Strategic Oscillation techniques presented in the Tabu theory (5) to realize such a coordination.
The two phases are sequenced as follow.
*T*he intensification phase (I)
*N*umber of satellites. At the beginning of the phase, the number of satellites included in the constellation belongs to a wide range; As the search progress, the number of required satellites becomes more and more precise and the tolerated number of satellite is centered on a mean value.
*N*umber of used classes. At the beginning of the phase, the number of usable classes is composed of the whole ODB. According to their value (computed from the simulation and exploration process) more and more classes are rejected (tabu state) as the intensification progresses.
The first consequence of this two evolution is a reduction of the neighborhood which facilitates the progression of the search at the high level stage.
*L*ow level optimization At the beginning of the phase, the low level stage is run with a few number of iteration and with low threshold in order to exit the optimization process very quickly. This parameters are regularly increased to make an more accurate low level optimization. With such evolution of the different parameters, we progressively transfer the effort (computes and time) from high level to low level.
*T*he Diversification phase (D)
After a long period of intensification, the search doesn't evolve anymore: the value of the best solution doesn't increase significantly; Visited configuration doesn't change. At this time, a diversification process is engage in order to explore other configuration types. The choice is making according to memory structures (frequency, recency) filled during the intensification phase. In fact, the less used classes are integrated to create a new configuration which will serve as a new departure for the high level stage.
Figure 3 summarizes all the developed notions.

6. Tests and applications

Principe

The validation of a hybrid algorithm is a challenging step. We have to analyze its behavior from different point of view: time computation, suitability

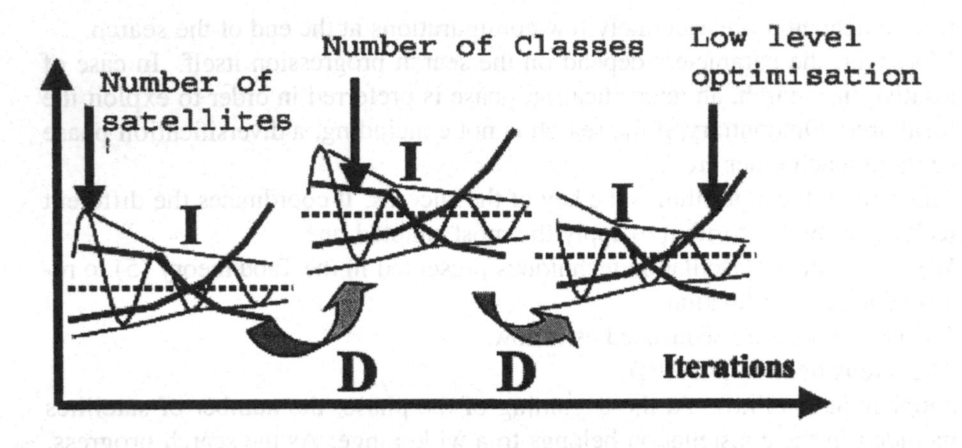

Figure 3. Strategic Oscillations

of returned solutions, pertinence and efficiency of each stage. To precisely analyze the behavior of the algorithm, we have to validate each stage separately before to make a global validation. In this paper, we are overall interested on optimization performances, more considerations about time are given at the end of the paper and in other papers (11,12,10,13).

Another relevant consideration (and difficulty) is that for most of the applications we don't know the best solution or simply a good solution. This remark could be applicable for many other complex systems.

To evaluate the low level optimization, we have to abstract the high level stage. So we set the configuration before to run the optimization process.

To evaluate the ability of this level to reach a good and/or the best solution we choose different kind of configurations: some configurations which contain a solution to the problem and some configurations which don't contain a solution.

The test of the high level stage must confirm the efficiency of the configuration navigation algorithm. We have to abstract the low level optimization. Then the corresponding algorithm must return the best solution within a configuration in order to inhibit the influence of the low level.

The corresponding algorithm (low and high level) is reduced to a combinatorial optimization process. After the separate validation, we have to integrate each stage to make a global validation of the algorithm. The main objective of this part is to finely tune the strategic oscillations to manage the low and high level trade-off during the search. In fact, time and iteration ratio change with the progression of the search.

The tests are also a way to analyze the behavior of the algorithm. So we should run scenarios we can check during their entire development.

Numerical stage validation

In our case, the validation has to be done for different application fields and we choose to apply the algorithm to telecommunication and navigation space system.
We present here a scenario for a continuous coverage of the [-70, +70] latitude band. We know an acceptable solution with 3 Geostationary (GEO) or 3 Medium Earth Orbit (MEO) satellites. So we both test the algorithm with 3 and 4 satellites in order to compare the results. There is no restarting strategy for this first test. Four initial solutions are injected to the algorithm and chosen as follow

- P0 : 3 Low Earth Orbit (LEO) satellites (4 LEO satellites)
- P1 : 3 Medium Earth Orbit (MEO) satellites (4 MEO satellites)
- P2 : 2 LEO + 1 MEO satellites (2 LE0 + 2 MEO satellites)
- P3 : 1 LEO + 1 MEO + 1 Geostationary (GEO) (2 LEO + 1 MEO + 1 GEO)

Table 1 present the results for the scenario 1. The stores values are the computation time and the evaluation of the best constellation reached during the search. Vmax indicates the best attended value. The test is made for three sets of relaxed parameters.
Results are satisfying because the best found solution for P1 condition with 3 satellites is acceptable for both relaxed parameters. In fact, the value of the best solution is about 95 percent of the best attended value. We could increase this value with a fine synchronization of the satellites but this is not the goal of the algorithm, we only want to know the configuration and global positioning of the satellites.
With 4 satellites, results are better but not concluding (the maximum is not reached). After a fine analyze, one of the satellites is redundant most of the time during the simulation. The three other satellites are correctly positioned but the forth one doesn't succeed to find a synchronized place.
The first statement concerning this low level algorithm is that during the first iterations the algorithm is able to find a value for the most influence parameters.
In fact, among the six parameters of a satellite, the altitude (a) and the inclination (i) are very determining for the quality of the solution.
This result is very promising because it allows considering a separated optimization which could dramatically reduce the time computation.

A second test has been realized with a restarting strategy, ?Table 2 stores the number of iterations to reach the optimum and the value of the optimum. Results are better than for previous test because the maximum is reaches with four satellites and reach 98.5 percents with three satellites.
The restarting strategy seems to efficiently bypass local optimum but the number of iterations is increasing consequently. For operational use, we have to take care of this to prevent exponential time consuming.

Metaheuristiques stage validation

To validate this stage, we create two kinds of ODB to ensure the return of a good constellation after each low level stage. Firstly, we voluntary choose an aberrant ODB composed of classes containing only one solution. The low level optimization is instantaneous.
Secondly, we built an ODB allowing a configuration containing a good solution. Corresponding classes allow several orbits and we have to precisely know the parameters values for the best solution in order to return it instantaneously.
We present now an application for navigation space systems. The goal is to recover the GPS constellation.
The main characteristics of this constellation are the doublet of satellites in each orbital plan and the symmetric distribution of the plane around the earth. The starting configuration is set to a Walker constellation with 18 satellites, in order to keep the symmetry.
The first result concerns the performances of the constellation and its design. Figure 4. presents the constellation design. We remark that a doublet is formed for each orbit plan. This result confirms the efficiency of such configuration for triangulation precision. The main difference between the returned constellation and the GPS constellation is the position of the doublet which is not symmetric. Global performances are quiet good because the criterion used to optimize is the availability of the constellation (that is to say the percentage of time the constellation is given a sufficient precision). The value of the returned constellation is about 99.89 percents and the value of the GPS constellation is about 99.999 percents. This values seems very close but the attended percentage should be very close to 100% for operational use.

The second result concerns the configuration evaluation. In fact, we define two ways to evaluate a constellation (a pre evaluation and a post evaluation). Figure 5. presents the evolution of this two values for a selected configuration during the high level search progression. Readers have to remind that the configuration values are linked to the class value which evolves at each new configuration exploration. The expression of these values are given in 13. We remark that the distance between the graphic stay small. The pre evaluation

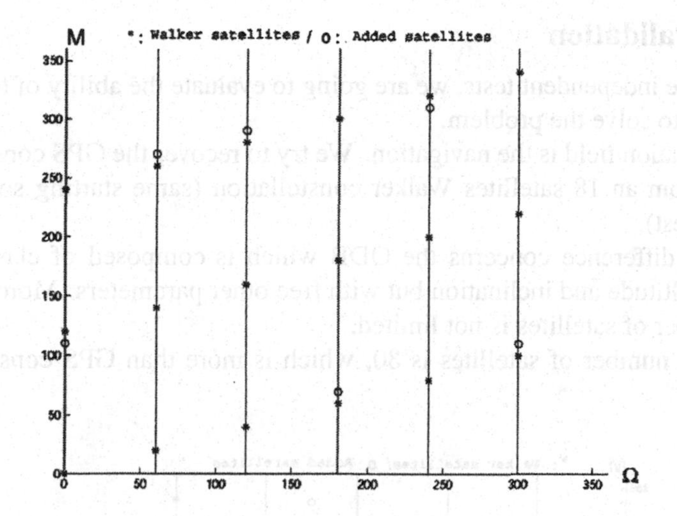

Figure 4. Constellation design

seems to be significant for a coarse estimation of the ability of a configuration to solve the problem.

Another interesting information concerning this test is that the high level

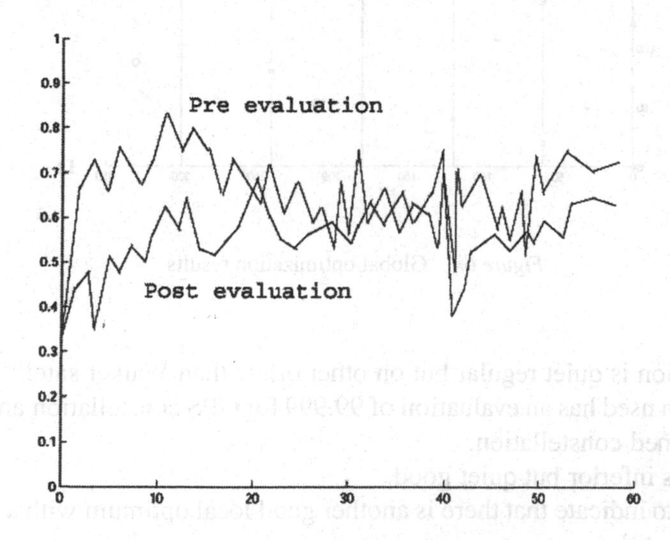

Figure 5. Pre and post configuration

algorithm only explores between 0.5 and 0.7 percents of the configurations. This allows considering a wide complex system optimization but should be decreased to treat the whole search space (6N parameters free).

Global validation

After the independent tests, we are going to evaluate the ability of the whole algorithm to solve the problem.
The application field is the navigation. We try to recover the GPS constellation starting from an 18 satellites Walker constellation (same starting solution as previous test).
The main difference concerns the ODB which is composed of classes with common altitude and inclination but with free other parameters. Moreover, the total number of satellites is not limited.

The total number of satellites is 30, which is more than GPS constellation.

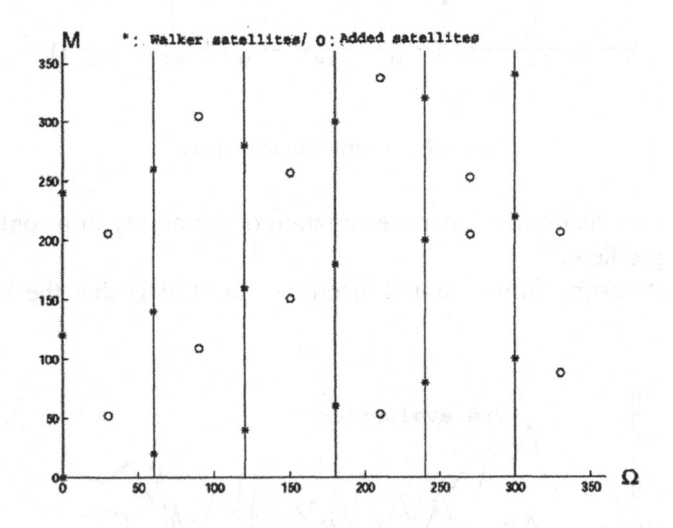

Figure 6. Global optimization results

The repartition is quiet regular but on other orbits than Walker satellites.
The criterion used has an evaluation of 99.999 for GPS constellation and 99.956 for the returned constellation.
This value is inferior but quiet good.
This seems to indicate that there is another good local optimum with a different satellite repartition.
We should also note that the GPS constellation is not necessary the best constellation for the proposed criterion.
On conclusion, the algorithm has been tested for both telecommunication and navigation systems. It presents satisfying characteristics in a directed way (precise orbit classes).

7. Operational use

This tool has to be used by engineers to help those designing constellations. The goal is not to deliver a standalone software but to give the possibility to model a constellation with certain constraints.

Using the tool as an expert system

According to the validation made in the previous section, the tool could be very efficient to find a sub-optimal solution. The efficiency is dependant on the ODB definition which is the way to drive and limit the search. Experts could use the ODB to introduce heterogeneous constraints such as altitude range, number of satellites or orbit inclination.
Experts could rapidly have a solution with a restricting ODB and so explore different kind of constellation. This first exploration allows establishing a global design of the constellation (minimum number of satellite, coarse range of altitude).
After this, an accurate search could be launched to determine precise parameters.
This way seems to be the best way to use the design tool.

Software development

The algorithm has been implemented in a high level computer science environment. A complete modeling of the system has been made with the UML language in order to produce a clean conception.
An object oriented language has been used to implement the algorithm and to develop a user friendly interface allowing to parameter the ODB and the whole algorithm (number of iterations, thresholds ...), to run the resolution and to visualize the results.
Figure 7 and Figure 8 present some snapshot of the main part of the interfaces. The main Window allows to parameter the Strategic Oscillation, the Tabu Search and the Steepest Descent. It displays the memory state, the best solution value, the number of ran iterations, the time spends. It also gives a link to visualize each satellite within the constellation by three ways: parameters values, 2D (Figure 9-a) and 3D (Figure 9-b) shape of the orbit. In a complex system, having different ways to visualize results (numerical, 2D, 3D, criterion) is important in order to analyze them.

How to speed up the search: towards high performance computing

Even if constellation satellite design is not directly dependant on tools performances or real time computing (this step is an upstream work not directly

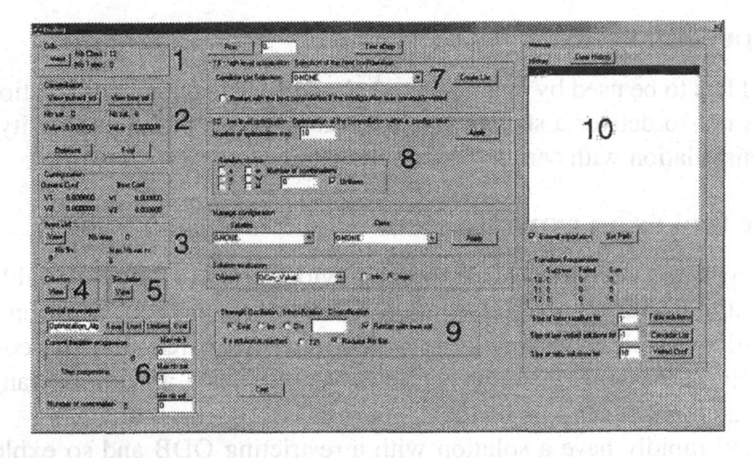

Figure 7. Main windows: algorithm

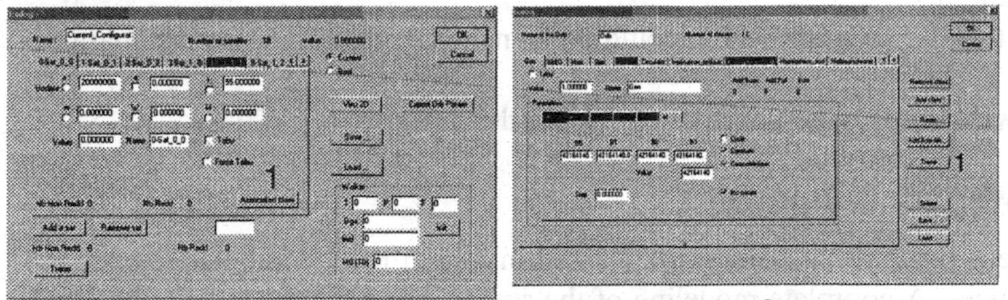

a) configuration definition

b) ODB Configuration

Figure 8. Sub window

linked to a production context), we have to consider an operational use case. Currently the tool is suitable for a quick solution outline or for a time consuming accurate solution.
As previously expressed, the main reasons are the time consuming evaluation criterion (sometimes several minutes for a single solution evaluation) due to a simulation process and the combinatorial exploration space engender by the parameters.
Regarding the system, parallelization seems to be the main way to accelerate this time consuming algorithm. In fact, we could consider the evaluation of each configuration of the neighborhood on several processors.
There aren't many parameters to transfer for each configuration to treat compared to the number of iterations to explore it. In fact, if we consider that the

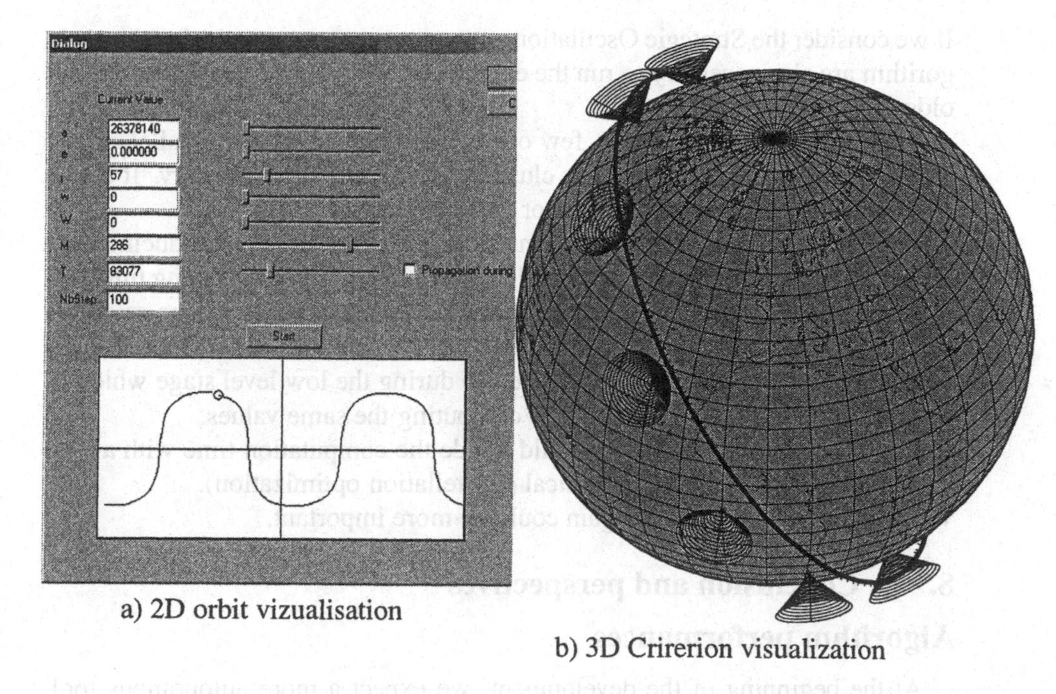

a) 2D orbit vizualisation

b) 3D Crirerion visualization

Figure 9. **Visualization**

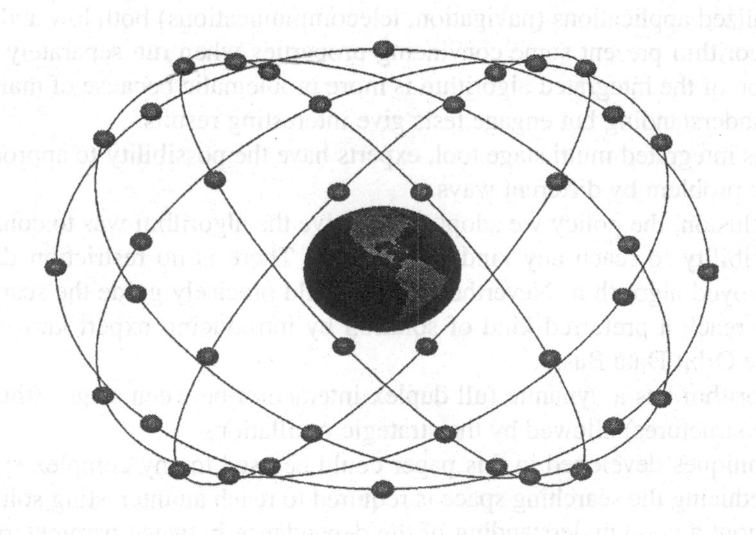

Figure 10. **3D constellation visualization**

ODB and the criterion are loaded on each processor, only N parameters are necessary to run a configuration exploration.

If we consider the Strategic Oscillation, parameters concerning the low level algorithm are also necessary to run the exploration (number of iterations, thresholds ...).
All those parameters cumulate, few octets are necessary and a parallel calculation is possible over a computer cluster with a distributed memory. It is not required to invest on a multiprocessor super calculator.
Other considerations, such as pre-calculation, allow considerably reducing time. In fact, when the criterion is evaluated, a simulation is run over a long time period and over a wide surface. When changing the parameters of one satellite, the whole computation is done again. So for a N satellites constellation, $\frac{N-1}{N}$ percent of the simulation is done twice. So during the low level stage which is a local search, we spend a lot of time computing the same values.
With this last consideration, we could divide the computation time with a factor between 10 and 20 (for a classical constellation optimization).
With a parallel approach, the gain could be more important.

8. Conclusion and perspectives

Algorithm performances

At the beginning of the development, we expect a more autonomous tool to design constellations. But operational considerations and accurate performances constraint us to propose a more interactive tool for expert designers.
For localized applications (navigation, telecommunications) both low and high level algorithm present some convincing properties when run separately. The validation of the integrated algorithm is more problematic because of main behavior understanding but engage tests give interesting results.
With this integrated multi stage tool, experts have the possibility to approach a complex problem by different ways.
On conclusion, the policy we adopt to conceive the algorithm was to conserve the possibility to reach any kind of solution. There is no restriction due to the employed algorithm. Nevertheless, we could precisely guide the search in order to reach a preferred kind of solution by introducing expert knowledge using the Orbit Data Bas.
The algorithm has a dynamic full duplex interaction between stages (through memory structures) allowed by the strategic oscillations.
The techniques developed in this paper could be used to any complex system where reducing the searching space is required to reach an interesting solution. But without a good understanding of the dependence between parameters and search space, the progression will be blind and not efficiency.

Extension to other real complex systems

The underlined advantages of this multi-stage algorithm are not specifically adapted to the constellation design problem.
In fact, the algorithm is customizable by many ways.
We do not have the presumption to propose a universal complex system optimization algorithm. Not at all. In this section, we only give indications to select and model problems having close characteristics.
The interaction between the algorithm and the application field (more precisely the problem to solve) is limited to the definition of ODB and evaluation function.
The ODB has been specifically created for the constellation design problem. As express in previous sections, this is an efficient way to drive the search and a specific effort has been made to design this knowledge database.
Such an effort must be done to adapt the ODB to the new challenging problem. We have to keep in mind, that the adaptation must conserve the characteristics which make the power of the ODB. That is to say, a way to intelligently limit the range of each parameter. In other words, each restriction must be justified with a physical correspondence. An elementary decomposition of the parameters range into uniform interval is not interesting and suitable for the algorithm. In fact, if changing the belonging class means changing the nature of the solution, the metaheuristic level of the algorithm could be able to extract significant structures from the explored solutions.
As example, the distribution of nodes in a network is a quiet similar problem. Likewise, the antenna positioning in cellular networks is close to current problem (24,25).
We could characterize an antenna (a node) with following parameters: cost, range (number of interface), bandwidth ...
Each kind of antenna (node) could be translated in terms of classes having physical signification.
The goal is to find the number of antenna (node), their characteristics (belonging class) and their position.
Each parameters of a class could be fixed or variable: fixed cost, or cost depends on bandwidth ...
Some of the parameters could be interpreted as constraints on localization: maximum altitude for an antenna, environment (high voltage wire proximity ...).

Acknowledgments

This work has been financed by Alcatel Space Industries (ASpI), located in Toulouse France. This industry is a world famous Space Industry. Contracts from many countries are handled each year by the DOS/AS department directed by Eric Lansard. This work was also due to a collaboration with the LIMA Laboratory of the ENSEEIHT, an engineer Computer Sciences School located in Toulouse, and particularly with Vincent Charvillat.

NSD = 1000			PSD	
Time/Val		a,i	a,i, e, w	a,i,e,w,W,M
Scenario 1	P0	16-15/280	32-18/315	47-55/362
3 Satellites	P1	16-38/1816	31-58/1756	48-17/1793
Vmax=2016	P2	16-15/784	32-22/724	48-31/637
	P3	16-56/1346	32-14/1278	48-29/1460
Scenario 1	P0	17-53/324	34-16/290	50-47/428
4 Satellites	P1	18-04/1912	34-22/1953	50-39/1887
Vmax=2016	P2	18-11/956	34-19/842	50-16/913
	P3	?	34-11/1314	50-29/1411

Table 1. Scenario 1 [-70, +70] latitude band coverage without restarting strategy

Vmax=2016			PSD	
		a,i	a,i, e, w	a,i,e,w,W,M
3 Satellites 1	P1	1700/1934	2100/1953	3050/1985
4 Satellites	P1	1467/2016	3300/1947	5400/2016

Table 2. Scenario 2 [-70, +70] latitude band coverage with restarting strategy

References

[1] Fuhan Tian. The Influence Of Crossover Operator On The Diversity In Genetic Algorithms. *IC-AI conference*, 2000.

[2] F. Glover Genetic Algorithms And Scatter Search : unsuspected Potentials, Statistics And Computing. , 1994.

[3] F. Glover Tabu Search Part I.. *ORSA Journal On Computing*, 1989

[4] F. Glover Tabu Search Part II. *ORSA Journal On Computing*, 1990.

[5] F. Glover and M. Laguna Tabu Search. *Kluwer Academic Publishers*, 1997.

[6] F. Glover and al. Finite Convergence Of Tabu Search. *MIC conference*, 2001.

[7] A.E. Smith and al. Genetic Optimization using a penalty function. , 1992.

[8] J. Grasmeyer Application of genetic algorithm with adaptive penalty functions to airfoil design. *48th International Astronautical Federation Congress*, 1997.

[9] J.T. Richardson and al. Some Guidelines For Genetic Algorithm With Penalty Functions. , 1993.

[10] E. Grandchamp and V. Charvillat Metaheuristics to Design Satellite Constellations. *Proceedings of MIC'2001*, 2001.

[11] E. Grandchamp and V. Charvillat Integrating Orbit Database And Metaheuristics To Design Satellite Constellation. *Proceedings of ICAI'2000*, pages 733-739, 2000

[12] E. Grandchamp and V. Charvillat Satellite Constellations Optimization With Metaheuristics. *Proceedings of Euro'2000*, pages 139, 2000.

[13] E. Grandchamp Some contribution to Satellite Constellation Optimization. *PhD Thesis*, 2001.

[14] D. B. Fogel An Introduction To Simulated Evolutionary Optimization. *IEEE*, 1994.

[15] D.G. Goldberg Genetic Algorithms In Search, Optimization And Machine Learning, Addison Wesley. , 1989.

[16] TZay Y. Young and Thomas W. Calvert Classification Estimation and Pattern Recognition. *Elsevier*,

[17] Zadeh and al. Fuzzy Sets. *Information and Control*, 1965

[18] Bandler and al. Fuzzy Power Sets and Fuzzy Implications Operators. *Fuzzy Sets and Systems*, 1980

[19] Dubois and al. A Class of Fuzzy Measures Based on Triangle Inequalities. *Int. J. Gen. Sys.*, 1983

[20] P.J Rousseeuw and al. Robust regression and outlier detection. *John Wiley and Sons*, 1987

[21] PJ. Huber Robust Statistics. *John Wiley and Sons*, 1981

[22] PT. Boggs et and al. Orthogonal Distance Regression. *Contempory Mathematics*, 1990

[23] Åke Björck Numerical Methods for Least Squares Problems. *SIAM*, 1996

[24] M. Vasquez and J-K. Hao "Logic-Constraint" Knapsack Formulation And A Tabu Algorithm For The Daily Photograph Scheduling Of An Earth Observation Satellite. *Journal Of Computational Optimization And Applications*, 2000

[25] M. Vasquez and J-K. Hao A Heuristic Approach For Antenna Positioning In Cellular Networks. *Journal Of Heuristics*, 2000

[26] CNES Techniques Et Technologies Des Véhicules Spatiaux. *Cépadues Editions*, 1989

[27] F. Zhang and al. Optimization Design Method and Its Application To Satellite Constellation System Design. , 2001

[28] A. Benhallam and al. Contribution Of Hybridization To The Performance Of Satellite Navigation Systems. , 1999

[29] P. Jannière and al. The Use Of Genetic Algorithms For The Design Of Future Satellite Navigation Systems. *ASTAIR*, 1994

[30] K. Kimura and al. Optimum Constellations Using Elliptical Inclined Synchronous Orbits For Mobile And Fixed Satellite Communications. , 1999

[31] M. Bello Mora and al ORION - A Constellation Mission Analysis Tool. *48th International Astronautical Federation Congress*, 1997

[32] F. Dufour and al. Constellation Design Optimization With A Dop Based Criterion. *14th Int. Symposium On Space Fligt Dynamics*, 1995

[33] T.A. Ely Satellite Constellation Design For Zonal Coverage Using Genetic Algorithms. *Space Flight Mechanics Meeting*, 1998

[34] H. Baranger and al. Global Optimization Of GPS Type Satellite Constellations. *42th International Astronautical Federation Congress*, 1991

[35] T. W. Beech and al. A Study Of Three Satellite Constellation Design Algorithms. , 1999

[36] C. Brochet and al. A Multiobjective Optimization Approach For The Design Of Walker constellation. *50th International Astronautical Federation Congress*, 1999

[37] N. Sultan and al. Effect Of Mobile Satellite System Design On Cost / Revenues Of Total Earth and Space Segments. *47th International Astronautical Federation Congress*, 1996

[38] F. Dufour and al. A Multistage Approach To Design And Optimize A Communication Satellite Constellaiton. *50th International Astronautical Federation Congress*, 1999

[39] Greistorfer Peter On The Algorithmic Design In Heuristic Search. *Euro conference*, 2000

[40] P. Hansen and al. Variable Neighborhood Search: Principles And Applications. *Les cahiers du GERAD*, 1998

[41] E. Lansard and al. Operational Availability. *48th International Astronautical Congress*, 1997

[42] Thomas Bäck and al. An Overview Of Evolutionary Algorithms For Parameter Optimization. *Evolutionary Computation*, 1993

[43] Z. Zhang Parameter Estimation Techniques: A Tutorial With Application To Conic Fitting. *Image And Vision Computing*, 1997

[44] Hastings and al. The Generalized Analysis Of Distributed Satellite Systems. , 1995

[45] J. Draim Optimization Of The Ellipso and Ellipso-2G Personal Communication Systems. , 1996

[46] R. G. Brown A Baseline GPS RAIM Scheme And A Note On The Equivalence Of Three RAIM Methods. *Journal Of The Institute Of Navigation*, 1992

[47] Bradford W. Parkinson GPS : theory and applications Volume I. , 1993

[48] D. Diekelman Design Guidelines For Post 2000 Constellations. *48th International Astronautical Federation Congress*, 1997

[49] M. Romay-Merino and al. Design Of High Performance And Cost Efficient Constellations For GNSS-2. *GNSS*, 1998

[50] E. Lansard and al. Global Design Of Satellite Constellations : A Multi-Criteria Performance Comparison Of Classical Walker Patterns And New Design Patterns. *47th International Astronautical Federation Congress*, 1996

[51] E. Frayssinhes Investigating New Satellite Constellation Geometries With Genetic Algorithms. *American Institute Of Aeronautics and Astronautics*, 1996

[52] F. Zwolska and al. Optimization Of A Navigation Constellation Design Based On Operational Availability Constraint. *GNSS*, 1998

[53] V. Martinot and al. Deployment And Maintenance Of Satellite Constellations. *49th International Astronautical Federation Congress*, 1998

[54] S. Abbondanza and al. Optinav : An Efficient Multi-Level Approach For The Design Of Navigation Constellations - Application To Galileo. *GNSS*, 1999

[55] Draim 9th DARPA Strategic Space Symposium. *GNSS*, 1983

[56] T. Lang and J.M. Hanson Orbital Constellations Which Minimize Revisit Time. *Astrodynamics Conference*, 1983

[57] J. G. Walker Circular Orbit Patterns providing whole earth coverage. *Royal Ear Craft Establishment Technical Report*, 1970

OPTIMAL DECISION MAKING FOR AIRLINE INVENTORY CONTROL

Ioana C. Bilegan[1,2], Sergio González-Rojo[1,3], Carlos A.N. Cosenza[4], Félix Mora-Camino[1,2]

[1]LAAS-CNRS, Toulouse, France; [2]ENAC/DGAC, Air Transportation Department, Toulouse, France; [3]ITCH, Chihuahua, Mexico; [4]COPPE/UFRJ, Rio de Janeiro, Brasil.

Abstract: This paper presents a market-reactive optimization approach to solve, on-line, the airline Revenue Management problem. The recursive demand forecasting method proposed, which makes use of geometric programming is described and the uncertainties related to the demand arrival process are taken into account within the Inventory Control module via the use of stochastic dynamic programming. A new backward recursive dynamic programming model is developed and implemented for different situations of fare-classes confinement and numerical results and performance assessments are obtained by computer simulations.

Key words: optimization; decision support systems; dynamic systems; market-reactive revenue management; demand forecasts; decision making under uncertainty; probabilistic model; geometric programming; dynamic programming.

1. INTRODUCTION

The aim of airline Revenue Management systems is to contribute to the efficiency of airline companies by maximizing their revenues obtained from selling the available seats on the flights they offer (fixed amounts of perishable services), based on reliable demand forecasts.

Following McGill and van Ryzin (1999), all revenue management systems can be considered to have four main components: forecasting, overbooking, seat inventory control and pricing. All those elements are important by themselves, but a natural hierarchy exists between them since, for example, the seat inventory control mechanism drives what optimization algorithms can be employed and the inputs needed by the optimization algorithm drive the demand model and what types of forecasts are needed.

This new management technique revolutionized airline industry and many other perishable-asset industries.

A reservation process has to be controlled and implemented within a complex decision support environment which should be able to make, on-line, the right decision with respect to any booking request received by the Computer Reservation System (CRS) at any point in time during the booking horizon.

In the proposed approach, the optimization module of the Revenue Management system works on-line, it gathers as input all the most recent updates provided by a demand forecasting function, as well as the present state of the reservations, to proceed with an optimization algorithm for the booking control process which treats new requests.

In order to take into account the highly stochastic nature of booking requests, forecasts should be updated with the latest information available. Therefore, a feedback control loop can be established between the Inventory Control module and the Demand Forecasting Updating process, leading to a market reactive Revenue Management system. The structure of the proposed system is then such as presented in Figure 1.

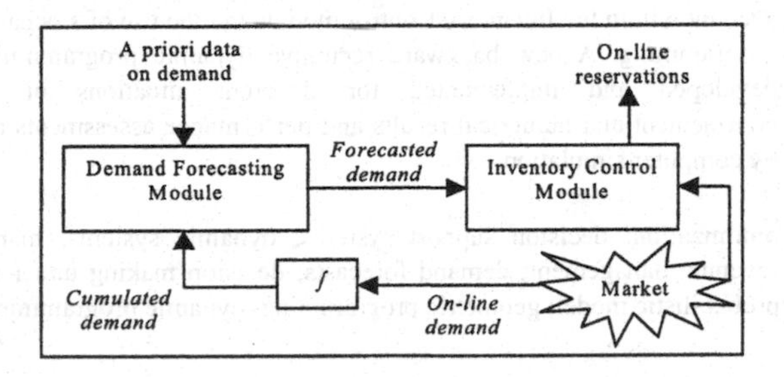

Figure 1. Simplified structure for airline RM systems.

2. PROBABILITIES UPDATING METHOD

The temporal (daily) dimension of the reservation process in the airline industry has to be taken into account for a realistic implementation of any optimization method. The Dynamic Programming technique, whose theoretical foundations were established in the early fifties, is used now in air transportation field with increased efficiency. Two main reasons can be quoted for that: first, the drastic increase in computational power of recent computers; second, the development of new forecasting methods which allow to estimate more accurately the probabilistic parameters describing the booking requests arrival process, which are needed by the Dynamic Programming algorithm.

In order to develop a stochastic Dynamic Programming approach to manage requests for travel occurring during predefined time periods, it is necessary to have at hand a representative probability distribution of demand by period, by fare-class and by order of arrival. Considering $p_{i,m}^n$ as the probability to have m demands for the class i during the decision period n, it is possible, for independent booking requests arrivals, to define the probability $\wp^n{}_{i,k}$ to have during decision period n, a kth demand for fare-class i:

$$\wp_{i,k}^n = \left(\sum_{l=0}^{K-k} \sum_{m_1+\cdots+m_l=k+l} \prod_{i=1}^{I} p_{i,m_i}^n \right) \cdot \frac{\sum_{m=0}^{K_i} m \cdot p_{i,m}^n}{\sum_{i=1}^{I} \sum_{m=0}^{K_i} m \cdot p_{i,m}^n} \tag{1}$$

with $K=K_1+K_2+\ldots+K_I$, K_i being the maximum number of demands for fare-class i during decision period n and I the number of fare-classes.

Relation (1) is based on the distribution of the $p^n{}_{i,m}$, which is the probability to have m demands for the class i during the decision period n. This distribution can be made available by a demand forecasting module and can be updated daily, based on the latest information (newly registered bookings) available to the Revenue Management system.

In the proposed approach, the updating process of the p_{jk}^n probabilities of upcoming demand makes use of a dual geometric programming formulation (Mora-Camino, 1978) for an optimization criterion of the information gain type (Jumarie, 1990):

$$\max \sum_{k=n+1}^{N} \sum_{j=0}^{J} \frac{p_{jk}^n}{N-n} \cdot \log\left(\frac{p_{jk}^{n-1}}{p_{jk}^n} \right) \tag{2}$$

under positivity (3), normality (4) and orthogonality (5,6) constraints:

$$\frac{p_{jk}^n}{N-n} \geq 0 \tag{3}$$

$$\sum_{k=n+1}^{N} \sum_{j=0}^{J} \frac{p_{jk}^n}{N-n} = 1 \tag{4}$$

$$\sum_{k=n+1}^{N}\sum_{j=0}^{J}\left(j-\frac{\delta_n}{N-n}\right)\cdot p_{jk}^{n}=0 \qquad (5)$$

$$\sum_{j=0}^{J}p_{j(n+1)}^{n}-\sum_{j=0}^{J}p_{jk}^{n}=0 \qquad (6)$$

where $k\in\{n+2, n+3, \ldots, N\}$, n is the decision period, j is the number of demands and $p^{n-1}{}_{jk}$ are the initial probability distributions, before updating. For more details and notations, see Bilegan, et al. (2001).

Once the problem is transformed into a non-constrained non-convex minimization problem, making use of the corresponding primal form of the geometric program, relation (7), genetic algorithms are used to solve it.

$$\min\phi(\vec{t})=\sum_{j=0}^{J}p_{j(n+1)}^{n-1}\cdot t_{n+1}^{j-\frac{\delta_n}{N-n}}\cdot\prod_{k=n+2}^{N}t_k+\sum_{k=n+2}^{N}\sum_{j=0}^{J}p_{jk}^{n-1}\cdot t_{n+1}^{j-\frac{\delta_n}{N-n}}\cdot t_k^{-1} \qquad (7)$$

with $t_k > 0$, for $k\in\{n+1, n+2, \ldots, N\}$.

The reasons of choosing genetic algorithms optimization technique (Goldberg, 1989) are given by the simplicity of encoding and the rapidness of global solution finding of the minimum of the non-convex continuous function $\phi(t)$, which is the objective of the associated primal geometric program.

Then, using geometric duality relations, the updated probability distributions will be finally computed and made available for further use, according with relations (8) and (9):

for $k=n+1$:

$$p_{j(n+1)}^{n^*}=\frac{(N-n)\cdot p_{j(n+1)}^{n-1}\cdot t_{n+1}^{*\,j-\frac{\delta_n}{N-n}}\cdot\prod_{k=n+2}^{N}t_k^*}{\sum_{j=0}^{J}p_{j(n+1)}^{n-1}\cdot t_{n+1}^{*\,j-\frac{\delta_n}{N-n}}\cdot\prod_{k=n+2}^{N}t_k^*+\sum_{k=n+2}^{N}\sum_{j=0}^{J}p_{jk}^{n-1}\cdot t_{n+1}^{*\,j-\frac{\delta_n}{N-n}}\cdot t_k^{*-1}} \qquad (8)$$

for $k=n+2$ à N :

$$p_{jk}^{n^*} = \frac{(N-n) \cdot p_{jk}^{n-1} \cdot t_{n+1}^{*\,j-\frac{\delta_n}{N-n}} \cdot t_k^{*^{-1}}}{\sum_{j=0}^{J} p_{j(n+1)}^{n-1} \cdot t_{n+1}^{*\,j-\frac{\delta_n}{N-n}} \cdot \prod_{k=n+2}^{N} t_k^* + \sum_{k=n+2}^{N} \sum_{j=0}^{J} p_{jk}^{n-1} \cdot t_{n+1}^{*\,j-\frac{\delta_n}{N-n}} \cdot t_k^{*^{-1}}} \quad (9)$$

3. DECISION MAKING USING DYNAMIC PROGRAMMING

For the inventory control part of the proposed approach, the probability distributions obtained with the forecast updating module described in Section 2 are used as input data for the implementation of a Dynamic Programming optimization module.

The technique of Dynamic Programming has been established by Bellman (1957) to cope with sequential dynamic optimization problems with applications in many different fields: Optimal Control (Bertsekas, 2000; Bertsekas, 2001), Operations Research (Hillier and Lieberman, 1967; Winston, 1994), Management Sciences (Fabrycky and Torgersen, 1966). Some authors (Lee and Hersh, 1993; Subramanian, 1995; Talluri and van Ryzin, 1998) have already made use of this technique, in a limited way, to cope with sequential decision process in airline Revenue Management.

In the previously published works about the application of Dynamic Programming to treat booking requests arrivals (Gerchak, et al., 1985), the following assumption was always made: in each decision period there is *at most one booking request* and the considered length of the successive decision periods must be such that this assumption holds. This approach has been accepted until recently on theoretical grounds, but it cannot be translated exactly into an on-line discrete decision process.

Therefore, in this communication, the proposed approach is such that the updating of demand forecasts is treated on a daily basis, turning the whole process of easy implementation. The main idea behind this proposal is the assumption that the forecaster has a better capability to perform accurate predictions over relatively large and fixed length time intervals (i.e. 24 hours, the natural daily time-frame of a booking process) rather than on variable and generally short-time decision periods.

The inventory control problem is solved here via a daily based Dynamic Programming model, developed for two distinct situations: physically unconfined and physically confined fare-classes. The following main *assumptions* have to be made for the two situations:

- the demand probability distributions for the different fare-classes are considered to be completely independent;
- only single leg flights are considered, cancellations and no-shows are not taken into account;

- go-shows are implicitly accounted for since bookings are permitted until the last moment before boarding closure; it is also implicitly considered that an estimate of the probability distributions for the "last day" demands (so, including the go-show probability distributions) are available;
- the whole capacity of the aircraft (C) forms a pool of seats available for reservations in all the fare-classes offered by the airliner on a single leg flight (no physically confined classes);
- the dynamic booking limits (changing with time/decision period and remaining available capacity) for each fare-class are made available by the solution of the expected revenue maximization problem;
- a decision period n, $n \in \{N, N\text{-}1, ..., 0\}$, lasts 24 hours (bookings by Internet can be performed at any time during the day);
- during each decision period n, a limited number of demand requests (min=0, max=K_i) for each of the fare-classes, $i \in \{1, 2 ..., I\}$, can be received by the reservation system;
- a booking request can be either accepted or rejected; if rejected, it is considered lost for the company (no recapture possibility is integrated in the model, the probability of buy-up is not quantified);

and according to the following *notations*:

- $n=N$ denotes the "first booking day" (the first decision period) of booking process for a given scheduled flight;
- $n=1$ denotes the "last booking day", i.e. the day of departure, before boarding closure of a given scheduled flight;
- $n=0$ denotes the period following the boarding closure before flight departure and during which no revenue can be achieved any more, regardless the number of seats still available; so the initial conditions for the recursive Dynamic Program will be indexed by $n=0$, as shown in relation (14).

3.1 Unconfined fare-classes

The first situation studied here considers the case in which the entire pool of seats of the aircraft cabin is available for bookings for all fare-classes offered on a flight-leg.

Let $\varphi_s^{n,k}$ be the maximum expected revenue to be obtained from the booking process when s seats are still available for booking and $(K\text{-}k)$ requests have been already made during the nth decision period. Then, the recursive expressions of the formulation of the backward Dynamic Programming model are given in relations (10) and (11):

$$\varphi_s^{n,k} = \left(1 - \sum_{l=0}^{K-k} \sum_{m_1+m_2+\cdots+m_I=k+l} \prod_{i=1}^{I} p_{i,m_i}^{n}\right) \cdot \varphi_s^{n,k-1} + \sum_{i=1}^{I} \wp_{i,k}^{n} \cdot \max\left\{F_i + \varphi_{s-1}^{n,k-1}, \varphi_s^{n,k-1}\right\} \tag{10}$$

$\forall$ n $\in$ {N, N-1, … , 1} and $\forall$ s $\in$ {0, 1, … , C}.

$$\varphi_s^{n,0} = \varphi_s^{n-1,K} \tag{11}$$

The decision criterion to accept or to reject an individual booking request is then given by:

$$F_i + \varphi_{s-1}^{n,k-1} \geq \varphi_s^{n,k-1} \tag{12}$$

This condition is based on the *expected revenue maximization* idea (Littlewood, 1972; Belobaba, 1989), which consists in accepting a booking request for a fare-class *i*, when *s* seats are still available for booking, only if the immediate revenue obtained from this (F_i) plus the maximum expected revenue to be obtained from the remaining available capacity (*s*-1) is greater or equal to the maximum expected revenue to be obtained if the actual booking request was refused (i.e. for future bookings, *s* seats would still be available).

In the case of group booking requests, this formula becomes:

$$q \cdot F_i + \varphi_{s-q}^{n,k-q} \geq \varphi_s^{n,k-q} \tag{13}$$

where *q* is the size of the group.

The initial conditions to compute the maximum expected revenue are such as:

$$\varphi_s^{0,k} = 0 \tag{14}$$

for $k \in \{1, \ldots, K\}$ and $s \in \{0, 1, \ldots, C\}$.

3.2 Physically confined fare-classes

The above model is extended here to the case where the aircraft cabin is divided in two different spaces, Business and Economy fare-classes, which have separate seat availability on the same flight-leg. Additional *assumptions* are necessary:

- a booking request for a seat in the Business Class can be either accepted, if there are seats available in the Business Class, or rejected (a business passenger cannot be accommodated on an economy seat – a lower standing);
- a booking request for a seat in the Economy Class can be accepted in the Economy Class, if there are seats available and can be also accepted in the Business Class if there are seats available in this class and no seats available in the Economy Class, or rejected (an economy passenger can be accommodated on a business seat – a higher standing);

and additional *notations* are adopted:

- the Business Class has a total capacity of C_b seats;
- σ is the number of seats still available for booking in Business Class;
- the Economy Class has a total capacity of C_e seats;
- s is the number of seats still available for booking in Economy Class;
- the total capacity of the aircraft is $C=C_b+C_e$;
- the business fare-classes are denoted by i, $i\in \{1, 2, \ldots, I_b\}$;
- the maximum number of booking requests that can arrive in Business Class during one decision period is given by $K_b= K_1+K_2+\ldots+K_{Ib}$;
- the economy fare-classes are denoted by i, $i\in\{I_b+1,\ I_b+2,\ \ldots, I_b+I_e\}$;
- the maximum number of booking requests that can arrive in Economy Class during one decision period is given by $K_e= K_{Ib+1}+K_{Ib+2}+\ldots+K_{Ib+Ie}$;
- in total, there are $I=I_b+I_e$ fare-classes offered on the aircraft.
- The formulation of a decision process based on Dynamic Programming for booking control is in this case more complex, since passengers can be assigned to two different pulls of seats.

Let $\varphi_{s,\sigma}{}^{n,k,j}$ be the maximum expected revenue when s seats are still available in the Economy Class, σ seats are still available in the Business Class, when $(K_e\text{-}k)$ Economy Class booking requests and $(K_b\text{-}j)$ Business Class booking requests have been already made during the nth decision period.

The recursive expressions of the formulation of the backward Dynamic Program in this second case are now given by:

$$\varphi_{s,\sigma}^{n,k,j} = \left(1 - \sum_{l=0}^{K_e-k} \sum_{m_{I_b+1}+\cdots+m_{I_b+I_e}=k+l} \prod_{i=1}^{I_b+I_e} p_{i,m_i}^n\right) \cdot \left(1 - \sum_{d=0}^{K_b-j} \sum_{m_1+\cdots+m_{I_b}=j+d} \prod_{i=1}^{I_b} p_{i,m_i}^n\right) \cdot \varphi_{s,\sigma}^{n,k-1,j-1} + \sum_{i=I_b+1}^{I_b+I_e} \wp_{i,k+j}^n \cdot \max\left\{F_i + \varphi_{s-1,\sigma}^{n,k-1,j}, F_i + \varphi_{s,\sigma-1}^{n,k-1,j}, \varphi_{s,\sigma}^{n,k-1,j}\right\} + \sum_{i=1}^{I_b} \wp_{i,k+j}^n \cdot \max\left\{F_i + \varphi_{s,\sigma-1}^{n,k,j-1}, \varphi_{s,\sigma}^{n,k,j-1}\right\} \quad (15)$$

with $n \in \{N, N\text{-}1, \ldots, 1\}$, $s \in \{0, 1, \ldots, C_e\}$, $\sigma \in \{0, 1, \ldots, C_b\}$, $k \in \{1, 2 \ldots, K_e\}$, $j \in \{1, 2, \ldots, K_b\}$

and by:

$$\varphi_{s,\sigma}^{n,0,0} = \varphi_{s,\sigma}^{n-1,K_e,K_b} \quad (16)$$

The initial conditions, adapted from the previous case, are such as:

$$\varphi_{s,\sigma}^{0,k,j} = 0 \quad (17)$$

The decision criterion to accept or reject an individual booking request in the Economy Class is in this case obtained by analyzing two possible situations.

If there are still some seats available in the Economy Class, the accepting condition established in the previous section can be adapted, from relation (12), for the pull of seats of the Economy Class; the $(K_e\text{-}k+1)^{\text{th}}$ booking request becomes an Economy Class reservation if:

$$F_i + \varphi_{s-1,\sigma}^{n,k-1,j} \geq \varphi_{s,\sigma}^{n,k-1,j} \quad (18)$$

To cope with the case where all the seats of the Economy Class are already booked while seats remain available in the Business Class, the accepting condition for an Economy Passenger on a Business Seat can be adapted from relation (12); then, the$(K_e\text{-}k+1)^{\text{th}}$ booking request becomes an Economy Class reservation if:

$$F_i + \varphi_{s,\sigma-1}^{n,k-1,j} \geq \varphi_{s,\sigma}^{n,k-1,j} \tag{19}$$

For a booking request in the Business Class, there is only one case that must be analyzed, since these booking requests cannot be accommodated elsewhere than in the pull of seats of the Business Class. If there are still some seats available in this class, the accepting condition established in the previous section can also be adapted, from relation (12), for the pull of seats of the Business Class. The $(K_b\text{-}j+1)^{\text{th}}$ booking request becomes a Business Class reservation if:

$$F_i + \varphi_{s,\sigma-1}^{n,k,j-1} \geq \varphi_{s,\sigma}^{n,k,j-1} \tag{20}$$

Otherwise, booking requests (either in Economy Class of Business Class) which cannot be satisfied by applying these conditions will be denied.

4. COMPUTATIONAL RESULTS

In this section a numerical example provides results obtained by the direct implementation of the backward Dynamic Programming model of the inventory control module, for physically unconfined fare-classes, presented in Section 3.

The simulation example consists in a single-leg flight from origin A to destination B, with three fare-classes ($F_1 > F_2 > F_3$). The available capacity, *4 days before departure*, is considered to be 20 seats. The maximum number of daily booking requests to come per fare-class is taken equal to 5, and thus the total maximum number of demands per decision period is considered to be 15.

The input data of the Dynamic Program consists of the probability distributions provided by the demand forecasting model which was described in Section 2, i.e. the direct application of relations (8) and (9) to solutions obtained using a classical (binary encoding) genetic algorithm optimization technique. For each of the three fare-classes, a different implementation, with different initial conditions for the genetic algorithm was used, such that for each fare-class different probability distributions, related to the value of the corresponding fare (the fare-class mean demand decreasing with F_i) were made available.

In Figure 2 (for fare-class 1), Figure 3 (for fare-class 2) and Figure 4 (for fare-class 3) the decision matrices obtained by using the proposed Dynamic Programming approach are presented. The example, which is representative for a high demand scenario (the total number of available seats is inferior to the total number of booking requests), corresponds to the decision period *n=2 days before departure*, where "*" means acceptance and "-" means denial of the corresponding booking request (when maximum 20 seats are still available for bookings and maximum 15 demands have already arrived to the CRS during the decision period of one day).

The performances of the decision making strategy obtained via the proposed Dynamic Programming model are evaluated by comparing the results of its implementation with the ones of the first come first served (FCFS) algorithm, applied to the same set of simulation data.

The comparative results presented in Figure 5 and Figure 6 demonstrate the superiority of the Dynamic Programming proposed approach.

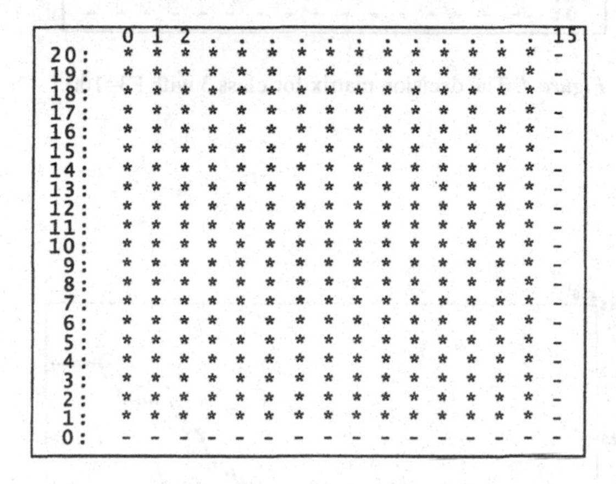

Figure 2. The decision matrix for class 1 with F1=500.

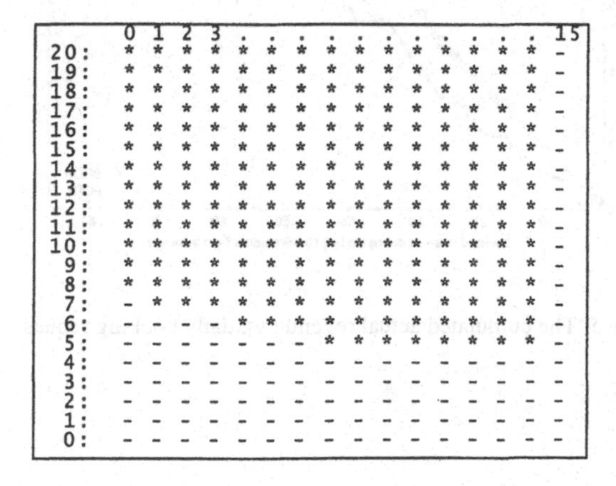

Figure 3. The decision matrix for class 2 with F2=200.

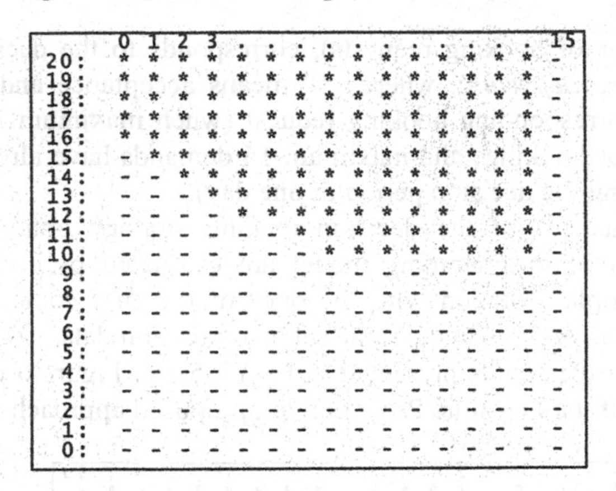

	0	1	2	3	.	.	.	.	.	.	.	.	.	.	.	15
20:	*	*	*	*	*	*	*	*	*	*	*	*	*	*	*	-
19:	*	*	*	*	*	*	*	*	*	*	*	*	*	*	*	-
18:	*	*	*	*	*	*	*	*	*	*	*	*	*	*	*	-
17:	*	*	*	*	*	*	*	*	*	*	*	*	*	*	*	-
16:	*	*	*	*	*	*	*	*	*	*	*	*	*	*	*	-
15:	-	*	*	*	*	*	*	*	*	*	*	*	*	*	*	-
14:	-	-	*	*	*	*	*	*	*	*	*	*	*	*	*	-
13:	-	-	-	*	*	*	*	*	*	*	*	*	*	*	*	-
12:	-	-	-	-	*	*	*	*	*	*	*	*	*	*	*	-
11:	-	-	-	-	-	-	*	*	*	*	*	*	*	*	*	-
10:	-	-	-	-	-	-	-	*	*	*	*	*	*	*	*	-
9:	-	-	-	-	-	-	-	-	-	-	-	-	-	-	-	-
8:	-	-	-	-	-	-	-	-	-	-	-	-	-	-	-	-
7:	-	-	-	-	-	-	-	-	-	-	-	-	-	-	-	-
6:	-	-	-	-	-	-	-	-	-	-	-	-	-	-	-	-
5:	-	-	-	-	-	-	-	-	-	-	-	-	-	-	-	-
4:	-	-	-	-	-	-	-	-	-	-	-	-	-	-	-	-
3:	-	-	-	-	-	-	-	-	-	-	-	-	-	-	-	-
2:	-	-	-	-	-	-	-	-	-	-	-	-	-	-	-	-
1:	-	-	-	-	-	-	-	-	-	-	-	-	-	-	-	-
0:	-	-	-	-	-	-	-	-	-	-	-	-	-	-	-	-

Figure 4. The decision matrix for class 3 with F3=100.

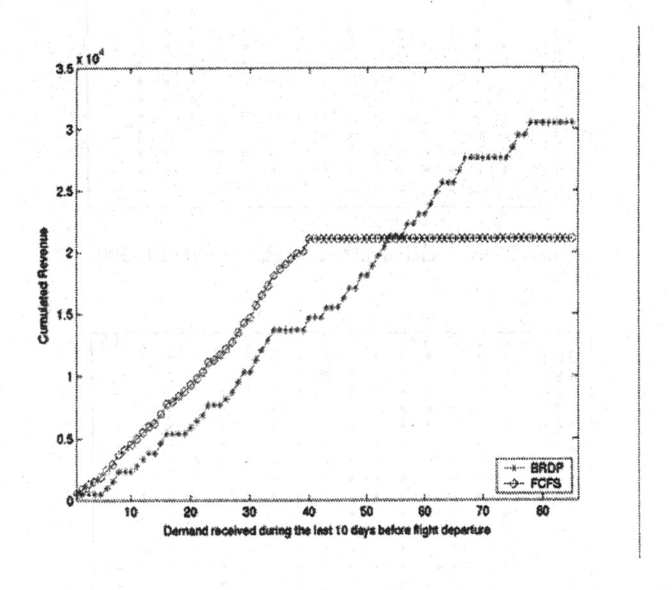

Figure 5. The cumulated actual revenue vs. daily booking requests.

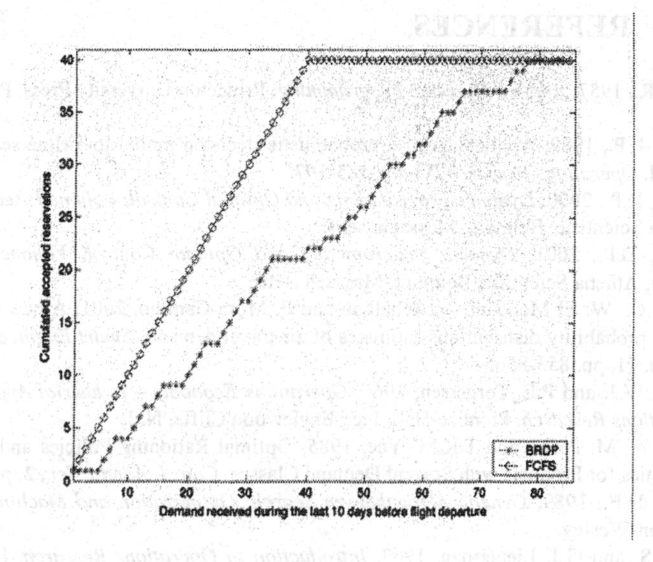

Figure 6. The cumulated accepted demand vs. daily booking requests

5. CONCLUSIONS

In this communication a new methodology based on dynamic programming has been proposed to establish an on-line airline inventory control process. This approach presents characteristics which turn it of real interest for practical utilization by airlines.

It is clear that, here again, the performance of the proposed control process depends on the quality of the available estimations of demand distributions. However, the recursive characteristic of the proposed optimisation method makes it fully compatible with an on-line demand estimation process which takes advantage of the newest available information.

Taking into account the simulation results obtained, presented in Section 4, good performances of the proposed methodology, when applied within a real system, are expected.

6. REFERENCES

Bellman, R., 1957, *Applied Dynamic Programming*. Princeton University Press, Princeton, N. J.

Belobaba, P.P., 1989, Application of a probabilistic decision model to airline seat inventory control. *Operations Research*, **37**, pp. 183-197.

Bertsekas, D.P., 2000, *Dynamic programming and Optimal Control, Volume I: second edition*, Athena Scientific, Belmont, Massachusetts.

Bertsekas, D.P., 2001, *Dynamic programming and Optimal Control, Volume II: second edition*, Athena Scientific, Belmont, Massachusetts.

Bilegan, I.C., W. El Moudani, A. Achaibou and F. Mora-Camino, 2001, A new approach to update probability distributions estimates of air travel demand. *Smart Engineering System Design*, **11**, pp. 853-856.

Fabrycky, W.J. and P.E. Torgersen, 1966, *Operations Economy – Industrial Applications of Operations Research*. Prentice-Hall, Inc., Englewood Cliffs, N. J.

Gerchak, Y., M. Parlar and T.K.M. Yee, 1985, Optimal Rationing Policies and Production Quantities for Products with Several Demand Classes. *Can. J. Admin. Sci.*, **2**, pp 161-176.

Goldberg, D.E., 1989, *Genetic Algorithms in Search, Optimization and Machine Learning*. Addison-Wesley.

Hillier, F.S. and G.J. Lieberman, 1967, *Introduction to Operations Research*. Holden-Day, Inc., San Francisco.

Jumarie, G.R.M., 1990, *Relative Information: Theories and Applications*. Springer Verlag.

Lee, T.C. and M. Hersh, 1993, A Model for Dynamic Airline Seat Inventory Control with Multiple Seat Bookings. *Transportation Science*, **27**, pp. 252-265.

Littlewood, K., 1972, Forecasting and Control of Passenger Bookings. *AGIFORS Proceedings*, **12**, pp. 95-117.

McGill, J.I. and G.J. van Ryzin, 1999, Revenue Management: Research Overview and Prospects. *Transportation Science*, **33**, pp. 233-256.

Mora-Camino, F., 1978, *Introduction à la Programmation Géométrique*. COPPE, Rio de Janeiro.

Subramanian, J., 1995, *Airline Yield Management and Computer Memory Management via Dynamic Programming*. Ph.D. Dissertation, University of North Carolina, Dept. of Operations Research, Chapel Hill, N. C.

Talluri K. and G.J. van Ryzin, 1998, An Analysis of Bid-Price Controls for Network Revenue Management. *Management Science*, **44**, pp. 1577-1593.

Winston, W.L., 1994, *Operations Research – Applications and Algorithms*: third edition. Duxbury Press, Belmont, California.

AUTOMATIC TEXT CLASSIFICATION USING AN ARTIFICIAL NEURAL NETWORK

Rodrigo Fernandes de Mello
Universidade de São Paulo
Instituto de Ciências Matemáticas e de Computação
Departamento de Ciências da Computação e Estatistica
Caixa Postal 668
13560-970 São Carlos – SP
mello@icmc.usp.br

Luciano José Senger
Universidade Estadual de Ponta Grossa
Departamento de Informática
Av. Carlos Cavalcanti, 4748
84030-900 Ponta Grossa – PR
ljsenger@icmc.usp.br

Laurence Tianruo Yang
Department of Computer Science
St. Francis Xavier University
Antigonish, NS, Canada
lyang@stfx.ca

Abstract The increasing volume of available documents in the World Wide Web has turned the document indexing and searching more and more complex. This issue has motivated the development of several researches in the text classification area. However, the techniques resulting from these researches require human intervention to choose the more adequate parameters to carry on the classification. Motivated by such limitation, this article presents a new model for the text automatic classification [1]. This model uses a self-organizing artificial neural network architecture, which does not require previous knowledge on the domains to be classified. The document features, in a word radical frequency format, are submitted to such architecture, what generates clusters with similar sets of documents. The model deals with stages of feature extraction, classification, labeling and indexing of documents for searching purposes. The classification stage, receives the radical frequency vectors, submit them to the ART-2A neural network that classifies them and stores the patterns in clusters, based on their similarity level. The labeling stage is responsible for extracting the significance level of each radical for each generated cluster. Such significances are used to index the documents, providing support to the next stage, which comprehends the document searching. The main contributions provided by the proposed model are: proposal for distance measures to automate the ρ vigilance parameter responsible for the classification quality, thus eliminating the need of human intervention on the parameterization process; proposal for a labeling algorithm that extracts the significance level of each word for each cluster generated by the neural network; and the creation of an automated classification methodology.

Keywords: text classifying, neural networks, adaptive resonance theory.

1. Introduction

Every day, millions of documents are added to the World Wide Web Pierre, 2000; Sinka and Corne, 2002. This increasing volume of available documents makes the searching operations more and more complex. Nowadays, the search engines use techniques based on the search of word sets within the Web documents. Such techniques spend a lot of computing processes as they require the indexing of all words contained on the documents and consider as of the same importance all the words contained there. Observing the complexity of the document search operations, several researches for text and document classification have been proposed Sinka and Corne, 2002; Pierre, 2000; He et al., 2003; Nigam et al., 2000; Blei et al., 2002. Out of these works, the following ones may be highlighted: M.P. Sinka e D.W. Cornei Sinka and Corne, 2002, J. He *et al.* He et al., 2003, K. Nigam *et al.* Nigam et al., 2000 and D.M. Blei *et al.* Blei et al., 2002. M.P. Sinka and D.W. Cornei Sinka and Corne, 2002 have proposed a pattern data set to accomplish the text classification studies. Such data set is based on *html* pages on 11 different subjects. Details on this data set are presented, in addition to the results that use the *k-means* algorithm to classify the documents. In order to use this algorithm it was necessary to count the word frequency on each document. For each document it was created a vector containing from 0.05% to 2% of the most frequent words. The created vector set was submitted to the *k-means* algorithm, which was responsible for calculating the minimum Euclidean distance among them. With the experiments, it could be proved that the documents average classification rate was above 50%, taking into account the previous classification as a pattern. On the experiments, the *k-means* algorithm was manually parameterized until the desired results were reached.

J. He *et al.* He et al., 2003 have proposed an alteration to the ART-2A artificial neural network to create a pre-defined number of clusters on the pattern classification. For this purpose, it was proposed a dynamic change on the ρ vigilance parameter, which may fit itself for the creation of a desired number of clusters. This technique is compared to the original ART-2A neural network,

SOM and to the *batch k-means* and *online k-means* clustering algorithms. The text classification results have proved that the change on the ART-2A neural network brings about, on certain instances, some improvements on the final results. The major contribution of this work is to allow the previous definition of the desired number of clusters by the end of a classification process.

K. Nigam *et al.* Nigam et al., 2000 have shown that the classification quality may be increased by means of a small set of documents previously labeled. It has been presenting a learning algorithm that uses labeled and unlabeled documents based on the combined techniques of *Expectation-Maximization* and a naive Bayes classifier. Before performing classifications with the algorithm, it is required to train the classifier with a subset of labeled documents. The knowledge obtained on this stage is then used to classify the remainder documents. Experiments have shown that errors on unlabeled documents decrease up to 30%.

D.M. Blei *et al.* Blei et al., 2002 have proposed a classification model that analyzes the global and local features of the documents. The global features comprehend the words and their frequencies; the local features are composed by the font size and color, italic, bold and other components related to the text format. Experiments have been carried out by using this model over a subset of documents for learning purposes and then on the remainder documents to classify them out. The experiment results have proved that classification errors are reduced in about 66%.

The previously presented work have shown good results but the choice of the best parameters is made by human intervention. For instance, M.P. Sinka and D.W. Cornei Sinka and Corne, 2002 have defined the *k* parameter, from *k-means* algorithm, fixed on 2, J. He *et al.* He et al., 2003 have defined a fixed number of clusters to be generated by the end of the classification process, what influences the neural network vigilance parameter and may deteriorate the classification quality. K. Nigam *et al.* Nigam et al., 2000 have shown that the classification quality may be enlarged by submitting a subset of labeled documents to the clustering algorithm. Nevertheless, this technique requires a previous human classification on part of the documents. D.M. Blei *et al.* Blei et al., 2002 have shown that the classification quality may be improved by considering the local features of each document. However, this technique requires previous training so that the model may learn the different manners to classify the remainder documents.

Observing the previously described limitations, this article presents a new model for automatic text classification. The documents may be stored in any format that may be transformed in pure text, such as *html*, *pdf*, *postscript*, *Microsoft Word*, *Microsoft Excel*, etc. This model is composed of the following stages: conversion of document format into pure text, stopwords removal, extraction of word radicals contained on the text, radical frequency counting, frequency submission to an ART-2A self-organizing artificial neural network which is responsible for classifying the document and extracting the radicals and their respective significances. Such radicals and significances are used to index the documents. From the keywords typed by the user, searches may be done on the radicals that index the documents, by arranging the result according to its significances. Thus, it is obtained,in a relevance order, the document set that best fits the user's request.

2. Model for Text Classification

This work has been motivated by the need to use human intervention to define the classification parameters on clustering algorithms such as *batch k-means*, *online k-means* and neural network architectures such as *SOM* and *ART*. This kind of parameterization has limited the automatic classification application over the unknown dominions. Such limitations have motivated the proposal for an automatic model for text classification, without the need of previous knowledge on the classification dominions and without human interference. The stages that make up this model are described on the following sections.

Feature extraction

The first stage of the automatic classification model is composed of the pre-processing and feature extraction, which provide subsidies for the classification start up. The steps that make it up are described as follows:

1 Conversion from document format to pure text - on this stage the documents are converted into pure text, so that the word occurrences may be counted on each document. The quantity of occurrences of a same word at a certain document, called α word frequency on the document, is used as the classifier entry.

2 Stopwords Removal- after the document is converted, the words that do not show relevant significance are removed. Examples of such words are: *of*, *the*, *in*, *and*, *or* etc. Thus, only the most relevant words are left to represent the document.

3 Extraction of word radicals contained in the text - with the main words of each document it is started the extraction of word radicals, responsible for attributing meaning to them. For instance, words such as *balancing*, *balance*, *balanced* are unified through the *balanc* radical.

4 Radical frequency counting - after the radicals are extracted, they are counted and the number of occurrences is stored at a distinct vector for each document. Such vectors contain the frequency of each word at a

certain document. Each Vt_i vector(where $i = 0, 1, 2, ..., n$) shows the same number of elements, where $|Vt_1| = k, |Vt_2| = k, ..., |Vt_n| = k$. Elements of a same index quantify the frequency of a same word. Thus, an α word is at the same j index of the $Vt_1, Vt_2, ..., Vt_n$ vectors.

Classification

The second stage is responsible for grouping and classifying the Vt_i vector sets (obtained on the first stage) by using an ART-2A Carpenter et al., 1991 self-organizing artificial neural network architecture.

The ART (*Adaptive Resonance Theory*) neural network family are self-organizing and non-supervised architectures that learn on stable representation codes in response to an arbitrary sequence of input patterns Carpenter and Grossberg, 1988; Carpenter and Grossberg, 1989. The ability of the ART-2A network family differs from the other self-organizing architectures as it allows the user to control the similarity degree among the patterns grouped at a same representation unit. This control allows the network to be sensitive to the differences existing on the input patterns and to be able to generate more or less classes in response to this control. Moreover, the learning on the networks ART is continuous: the network adapts itself to the incoming data, creating new processing units to learn the patterns, when required. Out of the different versions of networks ART, the architecture ART-2A may be highlighted as it allows the quick learning of the input patterns represented by continuous values. Because of its attractive features, such as noise filtering and good computing and classification performance, the neural ART network family has been used in several domains, such as to recognize Chinese characters Gan and Lua, 1992, interpretation of data originated on nuclear reactor sensors Whiteley and Davis, 1996; Whiteley and Davis, 1993; Keyvan and Rabelo, 1992, image processing Vlajic and Card, 2001, detection of earth mines Filippidis et al., 1999, treatment of satellite images Carpenter et al., 1997 and robots sensorial control Bachelder et al., 1993.

The ART 2A network architecture is composed of two main components: the attention and orientation systems (figure 1). The attention system is provided with an F_0 preprocessing layer, an F_1 input representation layer and with a F_2 class representation layer. The input and representation layers are interconnected through a set of adaptive weights called *bottom-up* ($F_1 \rightarrow F_2$) and *top-down* ($F_2 \rightarrow F_1$). The path from the neuron ith of layer F_1 to the neuron jth of the layer F_2 is represented by w_{ij}. Likewise, the neuron jth of layer F_2 is connected to ith of the layer F_1 through the adaptive weight w_{ji}. These weights multiply the signals that are sent among the neuron layers and are responsible for the storage of the knowledge obtained by the network. The interactions between the layers F_1 and F_2 are controlled by the orientation system, which uses a ρ vigilance parameter, and the way through which the weights are updated to obtain knowledge on the input patterns is defined by the training algorithm.

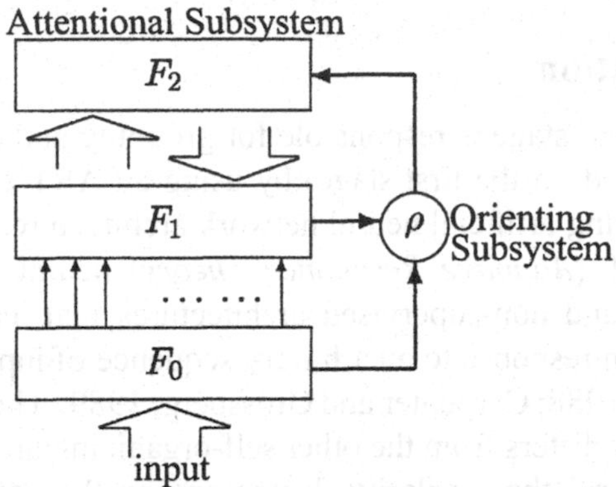

Figure 1. **ART 2A neural network basic architecture**

Training algorithm. The ART-2A dynamics are determined by the $\rho \in [0, 1]$ vigilance parameter and the $\beta \in [0, 1]$ learning rate. Initially, the output layer F_2 does not have any class. On our classification model, the input pattern I^0 is composed of a set of attributes associated to the process to be classified. The ART-2A training algorithm is composed of the following stages: preprocessing, activation, search, resonance or reset and adaptation.

- **Preprocessing:** this phase performs input normalization operations I^0:

$$I = \aleph(F_0(\aleph(Vt^0))) \tag{1}$$

where $\aleph$ and F_0 describe the following operations:

$$\aleph(x) \equiv \frac{x}{\|x\|} \equiv \frac{x}{\sum_{i=0}^{n} x_i^2}, \; F_0(x) = \begin{cases} x & \text{if } x > \theta \\ 0 & \text{otherwise} \end{cases} \tag{2}$$

Such operations perform the Euclidean normalization and noise filtering. The noise filtering, through the θ parameter, only makes sense if the main features of the input patterns, which lead to the creation of different classes, are represented *exclusively* on the highest values of the input components.

- **Activation:** this phase is responsible for sending out the incoming signals to the neurons of the representation layer F_2:

$$T_j = \begin{cases} Iw_{ij} & \text{if } j \text{ indexes a committed prototype} \\ \alpha \sum_j I_j & \text{otherwise} \end{cases} \quad (3)$$

  where T_j corresponds to the j neuron activation on the F_2 layer. Initially, all the neurons are marked as uncommitted and become committed when their weights are adapted to learn a certain input pattern. The α choice parameter defines the maximum depth of search for a fitting cluster. With $\alpha = 0$, value used in this work, all committed prototypes are checked before an uncommitted prototype is chosen as winner.

- **Search:** This phase is responsible for finding a candidate neuron to store the current pattern. The network competitive learning indicates the most activated neuron is the one chosen as candidate to represent the input pattern:

$$T_J = max \{T_j : \text{ for all } F_2 \text{ nodes}\} \quad (4)$$

- **Resonance or reset:** after selecting the most activated neuron, the *reset* condition is tested:

$$y_J > \rho \quad (5)$$

  If the inequality is real, the candidate neuron is chosen to store the pattern and the adaptation stage is initiated (resonance). If not, the winning neuron is inhibited and the searching stage is repeated (reset).

- **Adaptation:** this stage describes how the pattern will be learned by the network. Such stage comprehends the updating of the network weights for the J winning neuron, which, then, become committed:

$$w_{Ji}^{new} = \begin{cases} \aleph(\beta \aleph \Psi + (1-\beta) w_{Ji}^{old}) & \text{if } j \text{ indexes a committed prototype} \\ I & \text{otherwise} \end{cases} \quad (6)$$

$$\Psi_i \equiv \begin{cases} I_i & \text{if } w_{Ji}^{old} > \theta \\ 0 & \text{otherwise} \end{cases} \quad (7)$$

Table 1. ART-2A main parameters

Parameter	Description	Value example
m	number of input units	7
n	maximum number of representation units	15
θ	noise suppression parameter	$\theta = \frac{1}{\sqrt{m}}$
β	learning rate	$\beta = 0.7$
ρ	vigilance parameter	$\rho = 0.9$

The table 1 illustrates the examples of values for the ART-2A network parameters. The ρ vigilance value defines the quantity of classes that will be created by the network. The ρ value forms a circular decision boundary with a radius of $\sqrt{2(1-\rho)}$ around the weight vector of each category He et al., 2003. With $\rho = 0$, all input patterns are grouped at a same class. With $\rho = 1$, it is created a class for each input pattern presented to the network. The β learning rate defines the adaptation speed of the prototypes in response to the input patterns. The ART-2A should not be used with $\beta \cong 1$, as the prototypes tend to jump among the input patterns associated to a class instead of converging to the patterns average.

Each committed neuron of the F_2 layer defines a similar pattern group. The committed neuron set defines the classification generated by the network for the submitted values. As the input patterns over the process behavior are not previously labeled, it is required an algorithm to define a label to represent each class created by the network.

Labeling Algorithm. The labeling algorithm is built in accordance with the idea that the ART-2A network weights resemble the input patterns that have been learned by a certain neuron of the F_2 layer Senger et al., 2004. The ART-2A network weights are also called prototypes because they define the direction for the data grouping. The data normalization operations performed by the ART network allows all the vectors to be canonically normalized. Thus, only the angle formed among the prototypes is preserved. If the process monitoring values are initially normalized (see equation 2), such values do not differ too much in their magnitude. In addition, according to the rule for updating the ART network weights (see equation 7), the prototypes are also normalized. Thus, each attribute contribution can be obtained, based on its value on the weight vector. Such value represents the attribute significance for the local grouping chosen by the network.

After the data grouping is performed by the ART-2A network, a label is added to each neuron of the F_2 layer. For this purpose, a significance matrix is defined, which is composed of a SV_{ij} significance value set Ultsch, 1993. The significance matrix supports the decision about which components of the input vector are significant to label each committed neuron of the F_2 layer. The significance values are obtained directly from the ART-2A network weights, where the number of columns of the significance matrix is equal to the number of committed neurons of the F_2 layer, which represent the obtained classes and the number of lines is equal to the number of components of the input vector. For instance, a $SM = (SV_{ij})^{7\times4}$ significance matrix is obtained through a network that has 4 classes to represent a process described by an input vector with $m = 7$ components.

The labeling algorithm is illustrated by the Algorithm 1. In order to detect the most important attributes to describe the class, the significance values of the attributes are normalized in relation to the sum of the total significance values of a certain class, that is, the sum of the elements of the column. After such normalization is done, the column values are arranged in a decreasing manner and accumulated frequency of the significance values is calculated. For labeling the class, the set of the more significant attributes is selected until the accumulated frequency sum does not exceed a certain χ threshold. By the end of the algorithm execution there will be a C set of more relevant attributes, for each category created by the network, to label the class.

Algorithm 1 Labeling of the Classes obtained by the ART-2A network

1: defines the threshold value χ (p.e. $\chi = 55\%$) e
   m (input vector dimension)
2: create the significance matrix, one column for each class created by the ART-2A network
3: **for** each column created on the significance matrix **do**
4:     sum the significance values of each attribute
5:     normalize the significance values based on the sum
6:     calculate the distribution of the accumulated frequency
7:     arrange in a decreasing order the column attributes
8:     $sum := 0$
9:     $i := 1$
10:    $C := \emptyset$
11:    **while** ($sum \leq \chi$) **and** ($i \leq m$) **do**
12:        add the $attribute_i$ to the C set
13:        add the $attribute_i$ accumulated frequency to the sum variable
14:        $i := i + 1$
15:    **end while**
16:    label the class based on the attributes contained on C set
17: **end for**

The labeling stage obtains a C attribute set to represent the group. The elements of this set are afterwards used to index the documents.

Document Indexing and Search

The C attribute set, obtained through the labeling algorithm, is used to index the documents according to the significance of each word radical for one of the generated clusters.

After the indexing process, searches may be conducted from the keywords typed by the users. Out of these words are extracted the radicals, which are compared to the C attribute set that index the documents. By following this strategy, it may be obtained, in a relevance order, the document set that best fits the user's request.

3. Example on Text Classification

An example on text classification is presented on this section. Consider the table 2 as the word radical frequency counting of 10 distinct documents. The radical frequency set of each document is submitted to the ART-2A self-organizing neural network architecture, which is responsible for classifying it and creating clusters based on the document similarity.

Table 2. Radical frequency on Documents

Documents	finance	sport	biology	program	internet
1	10	5	7	0	1
2	0	1	15	10	7
3	2	0	19	5	3
4	3	7	8	8	5
5	1	9	3	0	10
6	8	15	2	9	11
7	2	1	10	12	12
8	1	0	8	2	1
9	10	0	7	7	5
10	0	12	0	9	3

The number of created clusters and stored patterns on each cluster varies according to the ρ vigilance parameter. The table 3 shows the quantity of clusters and the main word radicals, with their respective percentage significances to $\rho \in [0, 1]$.

On the table 3 it may be observed that high ρ values makes the neural network to generate a higher number of clusters. The column Identifiers Set and Significance of this table shows the identifier set and its significances for each created cluster. It may be noted that for the $\rho \in [0.05, 0.30]$ it is generated only a set of identifiers, that is, only a cluster is created. For $\rho \in [0.35, 0.40]$ are created two clusters and so on. The higher is the number of clusters, the more specific they get and may even reach the limit of containing only one pattern.

Table 3. Radical Significance for each Cluster

ρ	Clusters	Set of Identifiers and Significance (%)
0.05 to 0.30	1	{*finance* = 43.47, *biology* = 30.42}
0.35 to 0.40	2	{*finance* = 43.47, *biology* = 30.42}, {*sport* = 50.00, *program* = 37.50}
0.45	2	{*finance* = 43.47, *biology* = 30.42}, {*program* = 32.44, *internet* = 32.44}
0.50 to 0.65	3	{*finance* = 43.47, *biology* = 30.42}, {*biology* = 45.48, *program* = 30.28}, {*internet* = 43.51, *sport* = 39.12}
0.70 to 0.75	3	{*finance* = 43.47, *biology* = 30.42}, {*biology* = 45.48, *program* = 30.28}, {*internet* = 43.51, *sport* = 39.12}, {*sport* = 50.00, *program* = 37.50}
0.8	4	{*finance* = 43.47, *biology* = 30.42}, {*biology* = 45.48, *program* = 30.28}, {*internet* = 43.51, *sport* = 39.12}, {*finance* = 34.48, *biology* = 24.13}, {*sport* = 50.00, *program* = 37.50}
0.85	6	{*finance* = 43.47, *biology* = 30.42}, {*biology* = 45.48, *program* = 30.28}, {*internet* = 43.51, *sport* = 39.12}, {*sport* = 33.31, *internet* = 24.41}, {*finance* = 34.48, *biology* = 24.13}
0.9	7	{*finance* = 43.47, *biology* = 30.42}, {*biology* = 45.48, *program* = 30.28}, {*biology* = 25.80, *program* = 25.80, *sport* = 22.57}, {*internet* = 43.51, *sport* = 39.12}, {*sport* = 33.31, *internet* = 24.41}, {*finance* = 34.48, *biology* = 24.13}, {*sport* = 50.00, *program* = 37.50}
0.95	8	{*finance* = 43.47, *biology* = 30.42}, {*biology* = 45.48, *program* = 30.28}, {*biology* = 65.54}, {*biology* = 25.80, *program* = 25.80, *sport* = 22.57}, {*internet* = 43.51, *sport* = 39.12}, {*sport* = 33.31, *internet* = 24.41}, {*program* = 32.44, *internet* = 32.44}, {*finance* = 34.48, *biology* = 24.13}, {*sport* = 50.00, *program* = 37.50}

As the aim of the neural network is to classify similar patterns in a same cluster, extreme cases where there is only one pattern per cluster get undesirable. It appears, then, the issue of how to define the best value for ρ. In order to define its ideal value, this article proposes two performance measures that list the distances among patterns contained in each cluster, and the distance among centroids from distinct clusters He et al., 2003; Gokcay and Principe, 2000.

The first proposed measure is the intra-cluster distance defined by the equation 8. This measure, based on He et al., 2003, allows the quantification of the average distance among the patterns of a cluster and its c_j centroid. This measure calculates the average cosine of a $\theta_{i,j}$ angle among the $v_{i,j}$ vectors and the cluster c_j centroid, according to the cosine law Pappas, 1989.

$$Intra = \frac{\sum_{j=1}^{nc} \sum_{i=1}^{n} ||v_{i,j}|| * ||c_j||}{nc} \tag{8}$$

where nc is number of clusters in the ART-2A F_2 layer. The second measure, based on Gokcay and Principe, 2000, is the inter-cluster distance defined by the equation 9, which calculates the average distance among the centroids of clusters generated by the neural network. This equation, as well as the equation

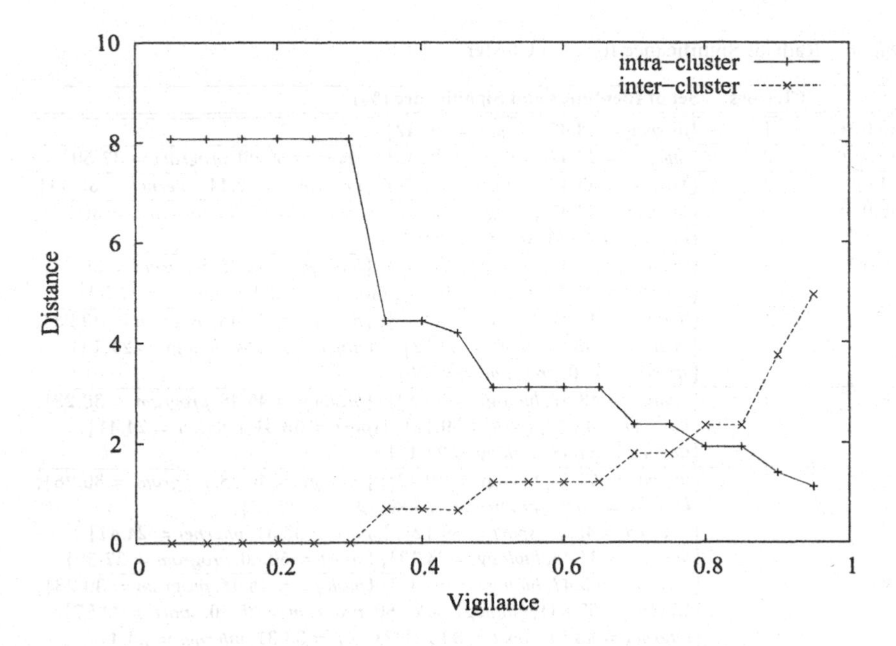

Figure 2. Experiment example: Variance of the intra and inter-cluster distance

8, also uses the cosine laws. Thus, the results of both equations are generated over a same scale, providing the definition of a point on the R^2 level where both functions get equal. This point defines the ideal ρ and, consequently, the number of clusters and patterns per cluster generated by the neural network.

$$Inter = \frac{\sum_{i=1}^{nc} \sum_{j=1}^{nc} ||c_i|| * ||c_j||}{nc} \tag{9}$$

The figure 2 shows the infra-cluster function and the inter-clusters for $\rho \in [0, 1]$. Through this figure it may be observed that the functions cross one another to $\rho \in [0.7, 0.75]$. On this point, it may be concluded that the intra-cluster distance is short enough to keep similar patterns at a same cluster, as well as the inter-cluster distance is long enough to separate the clusters created by the neural network. Low intra-cluster distance values implies on more and more similar patterns at the same cluster.

In order to validate the equations 8 and 9, experiments have been carried out using data sets where the ideal value for the vigilance parameter is known. The adopted data sets were Iris and Spanning Tree Blake and Merz, 1998 with the respective ideal vigilance values of 0.99 Vicentini, 2002; Vicentini and Romero, 2003 and 0.95 Fausett, 1994. The figures 3 and 4 show the results of the intra and inter-cluster distances to $\rho \in [0, 1]$. Observing the figure 3, it may be concluded that the ideal vigilance is from 0.98 to 0.99, which is a value very close to the Iris' ideal one. The figure 4 allows the conclusion that the ideal value for the Spanning Tree data set is around 0.95, what corresponds to the result obtained on Fausett, 1994.

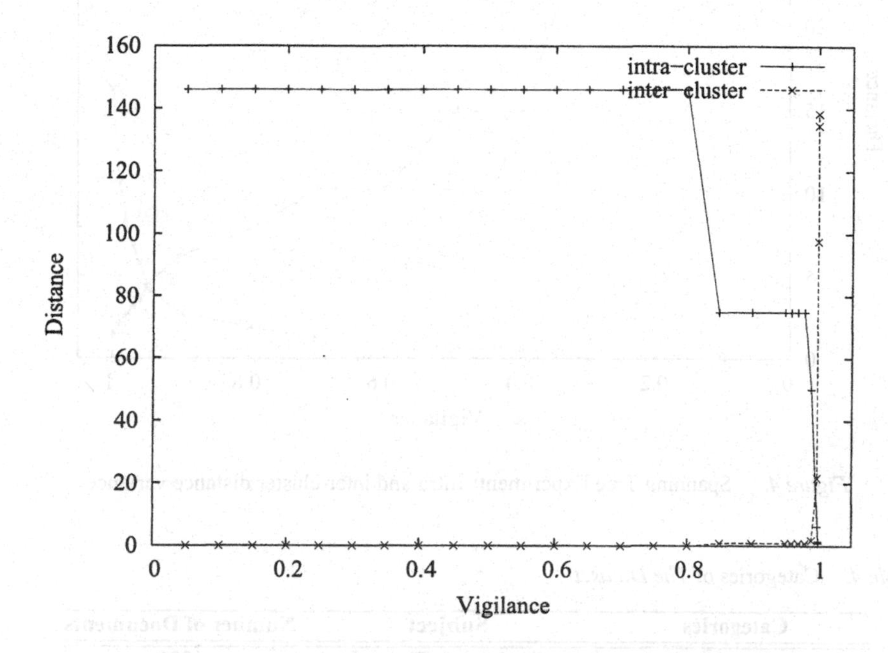

Figure 3. Iris Experiment: Intra and inter-cluster distance variance

The results of the experiments carried out on Iris and Spanning Tree data sets allows the validation of the results obtained through the distance equations that are used to automatically define the vigilance values, which were previously defined in a manual basis. Such equations are used to define the best vigilance values for the experiments presented as follows.

4. Experiments

In order to demonstrate the application of the proposed model, experiments have been carried out using the following data sets: *The Dataset* [2] and *20 Newsgroup* [3].

The first experiment was carried out using part of the *The Dataset* data set, which contains *html* pages ranked according to the table 4. 100 documents of the *Programming Languages - C/C++* type and other 100 from the *Science - Biology* type were evaluated. Out of these documents, the *tags html* [4] were removed.

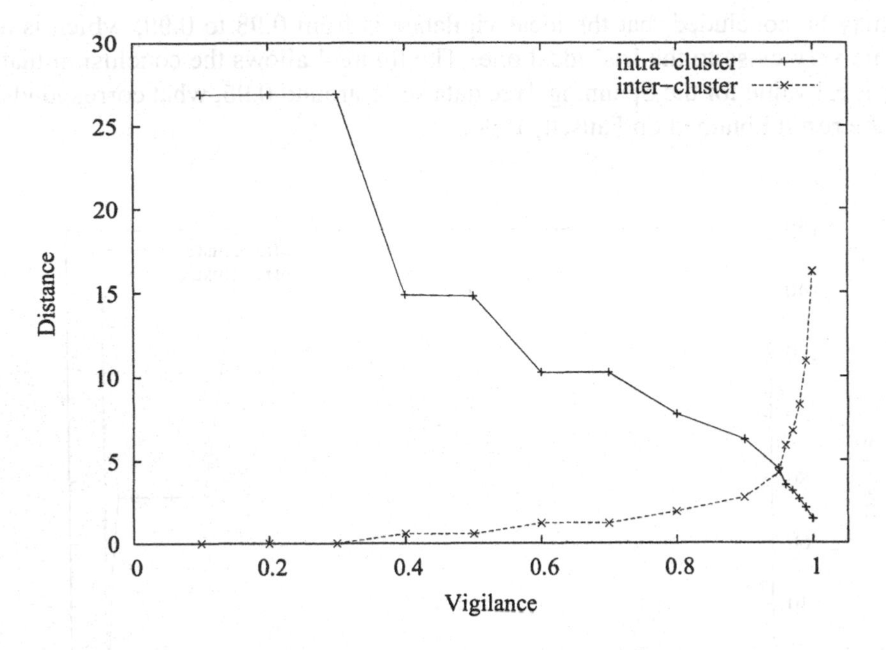

Figure 4. Spanning Tree Experiment: Intra and inter-cluster distance variance

Table 4. Categories of *The Dataset*

Categories	Subject	Number of Documents
`Commercial Banks`	Banking & Finance	1000
`Building Societies`	Banking & Finance	1000
`Insurance Angencies`	Banking & Finance	1000
`Java`	Programming Languages	1000
`C/C++`	Programming Languages	1000
`Visual Basic`	Programming Languages	1000
`Astronomy`	Science	1000
`Biology`	Science	1000
`Soccer`	Sport	1000
`Motor Sport`	Sport	1000
`General Sport`	Sport	1000

It may be noted that some *C/C++* and *Biology* documents were arranged on a same cluster. This fact has motivated the analysis of the file contents, which has proved that some texts of the *Biology* category had many keywords similar to the ones found on the C/C++ texts. In some cases there were programming language commands within the Biology texts. This allows the conclusion that the human classification made on these data sets does not deal with possible subject intersections contained on them. The classification results of the neural network for $\rho \in [0, 1]$ were evaluated by the equations 8 and 9, from which were generated the functions shown on figure 5. When making the regression of such functions, the equations on the inter and intra-cluster distances presented on table 5 were obtained. The equations were equalized in order to find out the point where both have the same distance. On this case, the ideal ρ is equal to 0.13568.

Found the ideal vigilance parameter, the document feature vectors were once again submitted to the neural network, which, then, classified them. Out of this classification were extracted the most relevant word radicals and their respective significances shown on table 6.

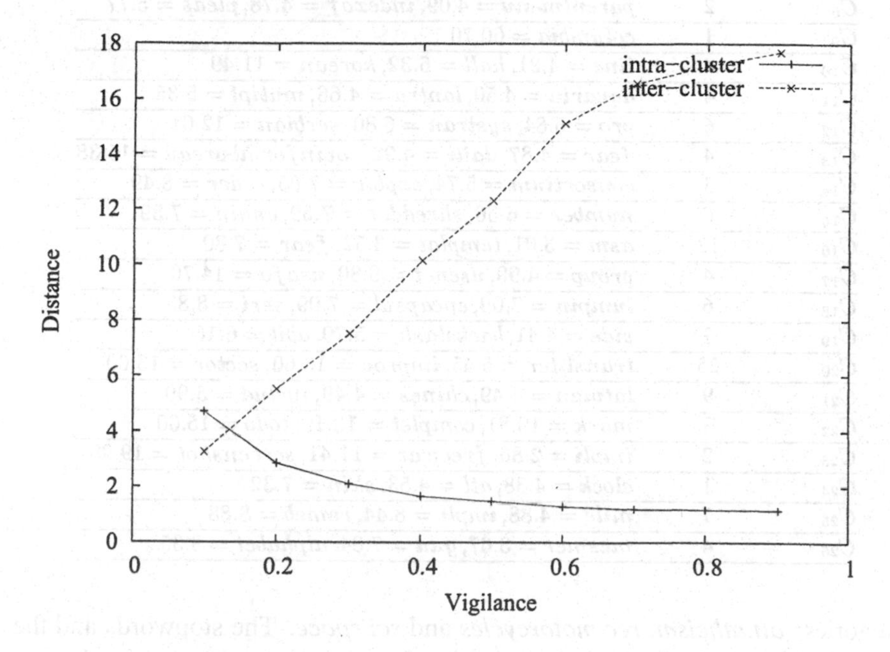

Figure 5. *The Dataset* Experiment: Intra and inter-cluster distance variance

Table 5. *The Dataset* Experiment: Equations of Distance

Distance	Equation	R^2
`Intra-cluster`	$-19.35x_3 + 38.34x_2 - 25.11x + 6.72$	0.989
`Inter-cluster`	$-35.70x_3 + 40.28x_2 + 10.02x + 1.96$	0.998

The second experiment has been carried out on part of the *20 Newsgroup* data set, which is composed of documents from a discussion list ranked according to the table 7. Out of these documents were removed *tags html* [5]. After this stage was completed, 300 documents were randomly distributed into three

Table 6. *The Dataset* Experiment: Radicals and Significance

Cluster	Patterns	Main Radicals e Significance (%)
C_0	15	$size = 5.08, deand = 5.16, e0001 = 7.07$
C_1	42	$site = 3.33, topic = 3.51, web = 6.31$
C_2	14	$lantra = 6.15, indexold = 8.22, free = 30.78$
C_3	2	$place = 3.35, vietnames = 12.98, glossari = 13.34$
C_4	2	$global = 44.86, yellow = 46.41$
C_5	5	$uni = 2.43, servic = 3.31, net = 7.72$
C_6	3	$lo = 5.99, chile = 6.45, espectador = 10.22$
C_7	7	$offic = 4.59, sharewar = 4.74, librari = 5.19$
C_8	2	$parentmenu = 4.09, indexof = 4.78, pleas = 5.77$
C_9	1	$columbia = 60.70$
C_{10}	4	$line = 4.21, hall = 6.32, korean = 11.49$
C_{11}	4	$aquariu = 4.50, lantra = 4.66, mutipl = 5.35$
C_{12}	6	$pro = 5.84, systran = 6.80, serbian = 12.01$
C_{13}	4	$fear = 4.87, daili = 4.92, sovinformbureau = 15.38$
C_{14}	3	$consortium = 5.74, explor = 7.65, coder = 8.49$
C_{15}	6	$number = 6.56, shredder = 7.59, engin = 7.59$
C_{16}	12	$asm = 3.01, templat = 3.52, fear = 7.80$
C_{17}	4	$group = 4.90, usenet = 9.80, usafa = 14.70$
C_{18}	6	$lumpin = 7.09, encapsul = 7.09, seri = 8.83$
C_{19}	7	$side = 4.41, backslash = 5.79, abil = 6.16$
C_{20}	25	$transliter = 5.45, improv = 13.60, sector = 13.60$
C_{21}	9	$latvian = 4.49, chines = 4.49, includ = 5.99$
C_{22}	3	$mark = 10.81, complet = 13.19, todo = 15.60$
C_{23}	2	$impli = 2.86, freewar = 11.41, screenshot = 19.98$
C_{24}	1	$clock = 4.38, atl = 4.53, skin = 7.32$
C_{25}	7	$mile = 4.88, impli = 8.44, remeb = 8.88$
C_{26}	4	$mesmer = 3.67, gun = 7.34, alphabet = 8.35$

categories: *alt.atheism*, *rec.motorcycles* and *sci.space*. The stopwords and the radicals were removed and their respective frequency counting was made.

The results of the intra and inter-cluster distances to $\rho \in [0, 1]$ are presented on the figure 6. This experiment included classification cases similar to the previous experiment, where documents from distinct categories were arranged in the same cluster. When the content of such documents was better analyzed, it was verified that they presented very similar subjects, although they were arranged in very distinct pre-defined categories.

Out of the results presented on figure 6, regressions were made and the equations of table 8 were obtained. By equalizing these equations it was possible to obtain the best vigilance parameter for the experiment being conducted, which was $\rho = 0.64432$.

After finding the ideal vigilance parameter, the patterns were once again submitted to the neural network, which then classified the documents. Out

Table 7. Categories of *20 Newsgroup*

Categories	Number of Documents
alt.atheism	1000
comp.graphics	1000
comp.os.ms-windows.misc	1000
comp.sys.ibm.pc.hardware	1000
comp.sys.mac.hardware	1000
comp.windows.x	1000
misc.forsale	1000
rec.autos	1000
rec.motorcycles	1000
rec.sport.baseball	1000
rec.sport.hockey	1000
sci.crypt	1000
sci.electronics	1000
sci.med	1000
sci.space	1000
soc.religion.christian	1000
talk.politics.guns	1000
talk.politics.mideast	1000
talk.politics.misc	1000
talk.religion.misc	1000

Table 8. *20 Newsgroup* Experiment: Equations of Distance

Distance	Equation	R^2
Intra-cluster	$9054.93x_4 - 21057.62x_3 + 17455.76x_2 - 6130.96x + 796.36$	0.983
Inter-cluster	$3497.84x_4 - 5497.00x_3 + 3019.84x_2 - 629.60x + 40.24$	0.995

of this classification were extracted the most relevant word radicals and their respective significances were presented on tables 9 and 10.

The classification generated by the ART-2A neural network architecture for the two experiments has provided a group of similar documents in a same cluster. Moreover, it was possible to extract the C attribute set (or word radicals) and its respective significances from each group. Such information allows the creation of a document database, which is indexed through the C attribute set. Over this database it is possible to make searches by means of keywords. As a result, documents are presented and they may be arranged according to their respective significance levels, based on the searching words defined by the users.

Table 9. *20 Newsgroup* Experiment: Radicals and Significance

Cluster	Patterns	Main Radicals and Significance (%)
C_0	309	$cleveland = 11.11, path = 12.12, usenet = 24.24$
C_1	5	$post = 15.80, cleveland = 16.62, research = 16.62$
C_2	5	$reston = 8.16, in = 10.20, howland = 14.28$
C_3	1	$jame = 7.51, srv = 10.04, messag = 12.54$
C_4	2	$juri = 8.76, felder = 11.66, da = 17.50$
C_5	4	$noc = 10.93, 3noc = 10.93, inc = 14.77$
C_6	2	$newsgroup = 10.20, harvard = 14.30, cmu = 26.56$
C_7	3	$mark = 7.70, line = 15.41, wisc = 33.42$
C_8	2	$da = 13.29, wisc = 17.72, reston = 22.16$
C_9	9	$mccullou = 11.19, edu = 14.95, cmu = 14.95$
C_{10}	18	$psuvm = 11.35, cmu = 17.02, watch = 23.53$
C_{11}	4	$don = 7.97, srv = 9.22, 10srv = 15.94$
C_{12}	4	$apr = 7.22, humbl = 7.85, harvard = 10.83$
C_{13}	6	$new = 11.10, cleveland = 13.26, quit = 17.97$
C_{14}	18	$research = 7.53, research = 22.84, odin = 28.51$
C_{15}	4	$research = 8.17, help = 13.63, phone = 15.11$
C_{16}	3	$messag = 13.60, inc = 15.23, ask = 22.67$
C_{17}	4	$technolog = 9.62, repli = 10.34, world = 18.11$
C_{18}	2	$sverdrup = 9.61, gmt = 10.37, don = 14.04$
C_{19}	3	$fun = 17.32, don = 38.95$
C_{20}	1	$post = 5.37, hopper3 = 17.84, watch = 21.81$
C_{21}	3	$atheism = 8.15, felder = 10.89, quit = 14.11$
C_{22}	5	$god = 7.98, cmu = 19.98, don = 27.67$
C_{23}	1	$cleveland = 6.47, obnoxi = 17.28, odin = 28.10$
C_{24}	3	$don = 8.32, drop = 8.32, respond = 31.68$
C_{25}	1	$david = 5.78, felder = 11.56, specif = 17.74$
C_{26}	3	$subject = 8.94, path = 12.52, specif = 18.94$
C_{27}	1	$atheism = 7.30, phone = 7.51, zue = 24.99$
C_{28}	23	$vishnu = 11.20, oath = 11.82, sverdrup = 22.41$
C_{29}	7	$affirm = 8.86, allah = 15.54, odin = 19.96$
C_{30}	7	$research = 5.00, technolog = 10.35, specif = 18.020$
C_{31}	1	$inc = 10.88, quit = 10.88, ask = 14.50$
C_{32}	5	$sverdrup = 9.07, gov = 9.56, couldn = 15.89$
C_{33}	3	$obnoxi = 7.34, sorri = 11.00, reston = 12.64$
C_{34}	2	$psuvm = 8.00, week = 10.52, couldn = 14.02$
C_{35}	2	$court = 9.42, reston = 10.80, testimoni = 14.27$
C_{36}	1	$server = 11.18, prejud = 11.18, spbach = 12.38$
C_{37}	6	$altern = 9.05, respond = 10.87, usenet = 20.90$
C_{38}	4	$alt = 15.34, follow = 17.11, atheism = 25.61$
C_{39}	3	$center = 8.26, edu = 14.43, help = 30.92$
C_{40}	1	$lewi = 8.48, reston = 9.57, sorri = 11.44$
C_{41}	75	$newsgroup = 12.98, dave = 16.36, respond = 21.84$
C_{42}	7	$harvard = 9.17, new = 9.17, affirm = 32.72$
C_{43}	19	$altern = 34.26, messag = 50.59$
C_{44}	44	$edu = 10.60, hopper3 = 21.21, spbach = 21.21$
C_{45}	1	$drop = 13.82, request = 20.74, line = 24.47$

Table 10. *20 Newsgroup* Experiment: Radicals and Significance

Cluster	Patterns	Main Radicals and Significance (%)
C_{46}	6	$nasa = 25.23, case = 41.12$
C_{47}	5	$newsgroup = 27.90, relev = 29.49$
C_{48}	11	$testimoni = 12.01, line = 20.40, respond = 24.02$
C_{49}	6	$cwru = 8.23, in = 16.43, prejud = 31.16$
C_{50}	9	$juri = 28.81, lewi = 44.80$
C_{51}	1	$theistic = 9.55, an = 20.32, srv = 33.82$
C_{52}	7	$host = 15.97, oath = 18.27, line = 22.79$
C_{53}	5	$part = 8.90, center = 17.53, psuvm = 26.31$
C_{54}	1	$recit = 14.30, specif = 21.45, gmt = 28.79$
C_{55}	15	$arizona = 8.03, lewi = 16.65, requir = 32.09$
C_{56}	154	$harvard = 14.69, organ = 22.06, request = 26.96$
C_{57}	7	$date = 29.27, ga = 30.51$
C_{58}	2	$mccullou = 13.46, david = 17.92, newel = 17.92$
C_{59}	6	$organ = 9.86, andrew = 12.41, date = 14.81$
C_{60}	1	$apr = 11.27, uug = 13.41, research = 20.66$
C_{61}	2	$recommend = 10.60, ga = 10.60, nntp = 18.44$
C_{62}	7	$help = 12.06, apr = 12.06, research = 18.08$
C_{63}	2	$monack = 9.53, organ = 13.30, subject = 17.75$
C_{64}	2	$help = 12.13, court = 14.55, atheism = 14.55$
C_{65}	2	$cmu = 10.99, 2cmu = 12.84, requir = 13.74$
C_{66}	1	$wisc = 8.75, line = 16.80, jesu = 17.47$
C_{67}	1	$felder = 19.17, world = 19.17, snake2 = 30.79$
C_{68}	1	$jame = 11.80, mccullou = 16.86, path = 35.37$
C_{69}	1	$fun = 11.29, newel = 16.92, center = 27.79$
C_{70}	2	$help = 8.67, andrew = 10.70, jame = 34.80$
C_{71}	1	$research = 10.14, mark = 10.50, center = 11.99$
C_{72}	5	$psu = 9.82, mark = 12.63, center = 14.43$

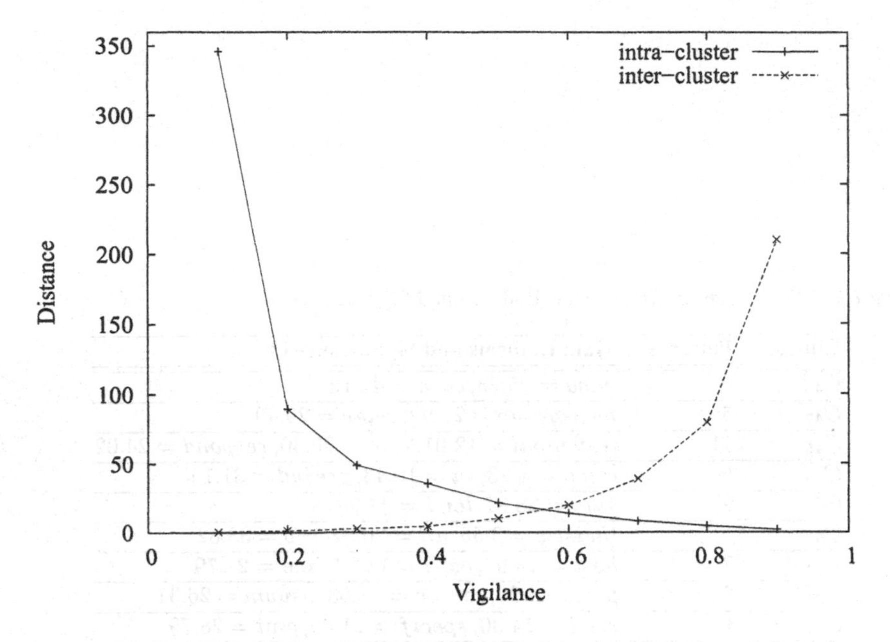

Figure 6. *20 Newsgroup* Experiment: Intra and inter-cluster distance variance

5. Conclusions

This article presents a new model for automatic text classification. This work has been motivated by the limitations of other classification techniques that require human intervention for the parameterization process Sinka and Corne, 2002; He et al., 2003; Nigam et al., 2000; Blei et al., 2002. The proposed model is composed of the following stages: feature extraction, classification, document labeling and indexing for searching purposes. The feature extraction stage comprehends the word radical frequency counting on the documents. Next stage uses an ART-2A neural network architecture to classify vectors with the word radical frequency on each document. The labeling stage is responsible for extracting the significance level of each word radical at each group generated by the neural network. Such significances are used to index documents, what supports the next stage that comprehends the document search. The main contributions of the proposed model are: proposal for distance measures to automate the ρ vigilance parameter, which is responsible for the classification quality, thus eliminating the human intervention on the parameterization process; proposal for a labeling algorithm that extracts the significance level of each word for each cluster generated by the neural network, and the creation of an automated classification methodology.

Acknowledgments

The authors thank to Capes and Fapesp Brazilian Foundations (under the process number 04/02411-9).

Notes

1. The software source codes to implement the model are available on

   http://www.icmc.usp.br/~mello/outr.html
2. Data set proposed by M.P. Sinka and D.W. Cornei Sinka and Corne, 2002
3. Available on http://www-2.cs.cmu.edu/afs/cs.cmu.edu/project/theo-20/www/data/news20.html
4. Available on http://www.icmc.usp.br/~mello/outr.html
5. Available on http://www.icmc.usp.br/~mello/outr.html

References

Bachelder, I., Waxman, A., and Seibert, M. (1993). A neural system for mobile robot visual place learning and recognition. In *Proceedings of the 5th Annual Symposium on Combinatorial Pattern Matching*, volume 807, pages 198–212, Berlin. Springer.

Blake, C. and Merz, C. (1998). UCI repository of machine learning databases.

Blei, D., Bagnell, J., and McCallum, A. (2002). Learning with scope, with application to information extraction and classification. In *Uncertainty in Artificial Intelligence: Proceedings of the Eighteenth Conference (UAI-2002)*, pages 53–60, San Francisco, CA. Morgan Kaufmann Publishers.

Carpenter, G. A., Gjaja, M. N., Gopal, S., and Woodcock, C. E. (1997). ART neural networks for remote sensing: Vegetation classification from lansat TM and terrain data. *IEEE Transactions on Geoscience and Remote Sensing*, 35(2):308–325.

Carpenter, G. A. and Grossberg, S. (1988). The ART of adaptive pattern recognition by a self-organizing neural network. *Computer*, 21:77–88.

Carpenter, G. A. and Grossberg, S. (1989). ART 2: Self-organization of Stable Category Recognition Codes for Analog Input Patterns. *Applied Optics*, 26(23):4919–4930.

Carpenter, G. A., Grossberg, S., and Rosen, D. B. (1991). ART 2-A: An Adaptive Resonance Algorithm for Rapid Category Learning and Recognition. *Neural Networks*, 4:4934–504.

Fausett, L. (1994). *Fundamentals of Neural Networks*. Prentice Hall.

Filippidis, A., Jain, L. C., and Lozo, P. (1999). Degree of familiarity ART2 in knowledge-based landmine detection. *IEEE Transactions on Neural Networks*, 10(1).

Gan, K. and Lua, K. (1992). Chinese character classification using adaptive resonance network. *Pattern Recognition*, 25:877–888.

Gokcay, E. and Principe, J. (2000). A new clustering evaluation function using renyi's information potential.

He, J., Tan, A.-H., and Tan, C.-L. (2003). Modified art 2a growing network capable of generating a fixed number of nodes. *IEEE Transactions on Neural Networks*. In press.

Keyvan, S. and Rabelo, L. C. (1992). Sensor signal analysis by neural networks for surveillance in nuclear reactors. *IEEE Transactions on nuclear science*, 39(2).

Nigam, K., McCallum, A. K., Thrun, S., and Mitchell, T. M. (2000). Text classification from labeled and unlabeled documents using EM. *Machine Learning*, 39(2/3):103–134.

Pappas, T. (1989). *The Joy of Mathematics*. Wide World Publishing.

Pierre, J. (2000). Practical issues for automated categorization of web pages.

Senger, L. J., de Mello, R. F., Santana, M. J., Santana, R. H. C., and Yang, L. T. (2004). An online approach for classifying and extracting application behavior on linux. In Yang, L. T. and Guo, M., editors, *High Performance Computing: Paradigm and Infrastructure*. John Wiley & Sons.

Sinka, M. and Corne, D. (2002). A large benchmark dataset for web document clustering.

Ultsch, A. (1993). Self-organising neural networks for monitoring and knowledge acquisition of a chemical process. In *Proceedings of ICANN-93*, pages 864–867.

Vicentini, J. F. (2002). Indexao e recuperao de informaes utilizando redes neurais da famlia art. Master's thesis, Instituto de Cincias Matemticas e de Computao da Universidade de So Paulo.

Vicentini, J. F. and Romero, R. A. F. (2003). Utilizacao de redes neurais da famlia art para indexao e recuperaqo de informaes. In *4th Congress of Logic Aplied to Technology*, pages 195–202.

Vlajic, N. and Card, H. C. (2001). Vector quantization of images using modified adaptive resonance algorithm for hierarchical clustering. *IEEE Transactions on Neural Network*, 12(5).

Whiteley, J. R. and Davis, J. F. (1993). Qualitative interpretation of sensor patterns. *IEEE Expert*, 8:54–63.

Whiteley, J. R. and Davis, J. F. (1996). Observations and problems applying ART2 for dynamic sensor pattern interpretation. *IEEE Transactions on Systems, Man and Cybernetics-Part A: Systems and Humans*, 26(4):423–437.